The Notion of Near Islands

RETHINKING THE ISLAND

The Rethinking the Island series seeks to unsettle assumptions by comprehensively investigating the range of topological and topographical characteristics that lie at the heart of the idea of "islandness."

Series Editors:

Elaine Stratford, Professor in the Institute for the Study of Social Change, University of Tasmania, Australia.

Godfrey Baldacchino, Professor of Sociology and Pro-Rector at the University of Malta, UNESCO Co-Chair in Island Studies and Sustainability.

Elizabeth McMahon, Associate Professor in the School of the Arts and Media, University of New South Wales, Australia.

Titles in the Series

Theorizing Literary Islands: The Island Trope in Contemporary Robinsonade Narratives, Ian Kinane

Island Genres, Genre Islands: Conceptualization and Representation in Popular Fiction, Ralph Crane and Lisa Fletcher

Postcolonial Nations, Islands, and Tourism: Reading Real and Imagined Spaces, Helen Kapstein

Caribbean Island Movements: Culebra's Trans-insularities, Carlo A Cubero

Poetry and Islands: Materiality and the Creative Imagination, Rajeev S. Patke

Contemporary Archipelagic Thinking: Towards New Comparative Methodologies and Disciplinary Formations, edited by Michelle Stephens and Yolanda Martínez-San Miguel

The Notion of Near Islands: The Croatian Archipelago, edited by Nenad Starc

The Islands and Atolls of the Maldives Archipelago, edited by Stefano Malatesta, Marcella Schmidt di Friedberg, Shahida Zubair and David Bowen (forthcoming)

An Introduction to Island Studies, James Randall (forthcoming)

Ecocriticism and the Island: Readings from the British-Irish Archipelago, Pippa Marland (forthcoming)

The Notion of Near Islands

The Croatian Archipelago

Edited by
Nenad Starc

ROWMAN & LITTLEFIELD
Lanham • Boulder • New York • London

Published by Rowman & Littlefield
An imprint of The Rowman & Littlefield Publishing Group, Inc.
4501 Forbes Boulevard, Suite 200, Lanham, Maryland 20706
www.rowman.com

6 Tinworth Street, London SE11 5AL, United Kingdom

British Library Cataloguing in Publication Information Available

Library of Congress Cataloging-in-Publication Data

Names: Starc, Nenad, editor.
Title: The notion of near islands : the Croatian archipelago / Edited by Nenad Starc.
Description: Lanham : Rowman & Littlefield Publishing Group, [2020] | Series: Rethinking the island | Includes bibliographical references and index. | Summary: "The main theme of this book is the exploration of the meanings of near islands, of the archipelago, as the chapters question and analyse the archipelago identity from different perspectives."—Provided by publisher.
Identifiers: LCCN 2020024779 (print) | LCCN 2020024780 (ebook) | ISBN 9781786610195 (cloth) | ISBN 9781786610201 (epub)
Subjects: LCSH: Archipelagoes—Croatia
Classification: LCC GB471 .N68 2020 (print) | LCC GB471 (ebook) | DDC 949.72/00942—dc23
LC record available at https://lccn.loc.gov/2020024779
LC ebook record available at https://lccn.loc.gov/2020024780

Contents

Acknowledgments

It would be difficult, if not impossible, to thank all those who supported us in our intention of writing a book about the Croatian islands. It is much easier to thank those who deserve it most, starting with Elizabeth McMahon, Elaine Stratford and Godfrey Baldacchino, the editors of the book series Rethinking the Island, who suggested that I gather the authors and edit what they wrote. Their comments and guidance have been invaluable. We have tried to follow them to the best of our ability. We are also grateful to Scarlet Furness and Gurdeep Mattu, editors at Rowman & Littlefield, for their support and advice.

In Croatia we received support from the islands and mainland. We would like to thank all those who encouraged us by realising that, after the publication of many books on individual Croatian islands, it was high time to write one about the archipelago. This book has been written by geographers, demographers, linguists, anthropologists, economists, and even a shipbuilding engineer and poet. Some are islanders, some are from the mainland, but they all have an island sensibility and readily agreed to write about not what is happening on the islands, but the islands themselves. This means they have knocked on the door of nissology. I hope they succeed in opening it. If they go on to write about the islands in their own terms, this book will have fulfilled its purpose.

The intention was to write a series of contributions which would be told as stories, understandable and interesting to the average reader, without sacrificing science. Thus, an archipelago of chapters about the archipelago began to form, but its readability needed testing. I sent draft outlines of the chapters to scientists, other people who deal with the islands in nongovernmental organisations (NGOs), and experts at the ministry charged with island development. I would like to thank Tajana Huzak and her associates, Ivana Hercegovac, Marija Mioč and Sanka Vučić, at the Ministry of Regional Development and

EU Funds. They carefully read everything I sent them, and their comments confirmed that the topic of the book and the approach were understandable to anyone interested in the islands. In terms of understanding, I am also grateful to my wife, Gordana, who put up with periods when I would work all day, then come home in the evening and continue working.

We wrote in Croatian, and Janet Berković tackled most of the translation work, mastering the jargon of several disciplines and wrestling with the bureaucratic prose favoured in official documents and legal acts. We are all grateful for this.

I must also include the University of Zadar, where four of our fourteen authors work. One of them, Vice-Rector Josip Faričić, enthusiastically welcomed the idea for the book and directed me towards other authors. The university also provided financial support, for which I am particularly grateful.

Finally, I must mention the Institute of Economics in Zagreb, where I spent my working life and remain active as an emeritus. They allowed me to use my old study and the available technical support. As I am also trying to open the door of nissology and do not intend to stop anytime soon, I am particularly grateful for this.

Nenad Starc

Introduction

Nenad Starc

It should be easy enough to define an island: a body of land surrounded by water. The definition is clear and appealing, and we do not tend to ask if it covers too much or if it is practical enough. Bodies of land in the water (most often we think of the sea) come in all sizes, however, and can be also referred to as rocks, reefs and islets. Only when it increases to 1 km^2 or more can it indeed be considered an island. Biologists might add that a proper island has both marine and land life, soil, flora, and fauna. The Japanese agree and write 島 (*shima*), which is translated, "a bird on the grass on the sea," or sometimes "bird on the mountains." We can only hope that biologists (at least in the West) will acknowledge that there is a place in science for the odd haiku. Or even runes. The Swedes write *Ö* for an island and call them Värmdö, Lidingö and Utö. Sometimes, even bureaucrats come in with surprisingly picturesque contributions. In the 1861 Scottish census, for instance, islands were recognised as places where a sheep (at least one) can find enough food to survive. It seems that defining an island appears to range from the mundane to the evocative, and as long as it involves two media, land versus sea, there is no confusion (Baldacchino 2006; Stratford 2003; Royle 2007). Some may consider plains as bodies of land surrounded by mountains, while never defining them as an island. After Herodotus, who called the Saharan oases an archipelago, few people have discovered islands in the desert (Morris 2010), and hardly anyone has viewed lakes as islands of water surrounded by land (Matvejević 1987). The definiens of being surrounded by water is reserved only for islands.

Language reveals cultural imprints when it comes to rethinking the island concept (Ronström 2009; Royle 2007). Seafaring nations understand the diversity of islands and use a variety of names. English is especially rich in this respect. The word *land* appears in such words as island and mainland. A *body of land, landmass, a piece of land, a tract of land* against the medium of water is self-explanatory and bears no conflicting connotations. Peoples who moved to the coast accompanied with an inland-built language tended to use the words they had rather than invent new ones. When the Croats arrived on the Adriatic coast thirteen centuries ago, they brought their word for island with them. They had observed how rivers would flow around a piece of the mainland in mid-channel and called it an *o* (around) *tok* (flow). Accordingly, larger pieces of land in the sea throughout the Croatian side of the Adriatic were named *otok*, and smaller ones were hardly scrutinised. It is perhaps confusing to today's islanders that there are 9 towns in inland Croatia, 2 towns in Bosnia and Hercegovina, and 2 towns in Slovenia named Otok.

Even on modern nautical charts, we still find points that have been called reefs for centuries, although they remain visible at high and low tide. It appears the inland experience was useful, although limited. Having found a name, Croats still had to define these many bodies of land observed from the shore and decided an *otok* was "a piece of mainland" surrounded by sea. Croatian geography and nautical textbooks, and the Croatian Islands Act, contain this definition. Christened and established according to the river and mainland experiences, the Croatian islands are locked in the notion that they are extensions of the mainland. Whether the definiens suffices is hardly questioned; the difference between islands and the mainland is swept under the carpet.

The Latin words *insula* and *insular* were used more commonly on the mainland, not only in continentally bound Croatia. Ancient Roman towns were full of *insulae*, and perhaps our insular cortex helps us to consider some parts of the mainland insularised regardless of their distance from the sea. The Finns have attended to this. Their Islands Development Act (1981) states, "An island is a body of land surrounded by the sea or a lake," but also "any mainland area in which such conditions prevail." This example is perhaps limited, as legislation is a particularly fluid medium. The Croatian Islands Act (1999) covers the topic of islands in 56 articles and includes one peninsula, without ever saying what an island is. It appeared so evident that no one thought of defining it. The Japanese legislator had no such issue; the *kanji* 島 used in the Island Act (1952) is both the name and the definition.

Leaving aside the medium of water for a moment and concentrating on what it means to be separate and surrounded, many things become islands; insularised regions and subregions, and insular art; green islands in managing the urban waste; pedestrian islands in city traffic; and islands in the middle of the kitchen. Just for pastime, let us remember that long, long ago, there was Pangea, the one and only piece of land on Earth, which was undeniably surrounded by the sea on all sides. When it separated into Gondwana and

Laurasia, it begs the question of which was the mainland and which was the island? This sort of thinking, which apparently leads nowhere, might take us to Australia, which is certainly a continent but is also considered an island (McMahon 2010), or to Greenland, a 2 million km^2 island that is never thought of as a continent (Grydehøj 2018). Moreover, it is often dispatched to a footnote where it is separated from other world islands since its size upsets statistics and disrupts indicators.

The usual definitions, no matter the language, are hardly adequate but inspire curiosity. The difficulty relates to a host of issues as they cover too much (Lewis and Wigen 1997; Hay 2006; Baldacchino 2007; Baldacchino 2013): open-sea islands and those close to the mainland, between which it is hard to distinguish (Deleuze 2004); islands joined to the mainland by bridges, making them part of the mainland, albeit still surrounded by water; islands that have been turned into peninsulas by filling in channels but are still insularised in many ways; peninsulas that have become islands overnight due to dredging and islands that have emerged from the sea by being heaped up artificially. There are also urban archipelagos and even hyperinsularised and hypoinsularised islands. We should not omit parts of the mainland simply declared as islands, not only in Finland and Croatia. Even the most straightforward division into large and small islands can spark debate. The United Nations Educational, Scientific, and Cultural Organization (UNESCO) states that islands with surface areas less than 10,000 km^2 and fewer than 500,000 inhabitants should be considered small (Hess 1986). The European Small Islands Federation (ESIN) only deals with islands with surface areas of less than 100 km^2 and fewer than 100 inhabitants, while EUROSTAT determines that an island is small if NUTS (nomenclature des unités territoriales statistiques—nomenclature of territorial units for statistics) classification cannot capture it. Furthermore, until as recently as 2017, EUROSTAT considered Malta and Cyprus mainland areas because their capitals were on the island.

Another way of rethinking definition is to reverse perspective so that the mainland becomes the secondary term or point of comparison enabling us to turn to the island in its own terms. Travel writers have reported on islands in this way for centuries, and writers of fiction often use them as a mise-en-scène, showing that they have seen something not found on the mainland. Island literature has grown into a tried and tested genre, often concerning such themes as buried treasure, being utterly alone in the world, utopias and, in particular, dystopias, as with the reign of the Lord of the Flies. Finally, scientists have turned to the islands too and discovered that much of what happens on both islands and the mainland is better preserved, more noticeable, free from surrounding influences and, in a word, more suitable for research on the islands. From Darwin, who established his theory of evolution on the

Galapagos archipelago, to the present-day naturalists and social scientists, islands have proven to be favorable research destinations. It has often been heard that islands are laboratories where results come out better than on the mainland. The first documents that went a step further have only appeared recently, in the 1990s. It was simply proposed that islands should be addressed in their own terms (McCall 1994).

The proposal for an island-centred analysis fell on fertile ground and marked the beginning of a new scientific discipline. In search of a name, its founders returned to Ancient Greece, as was only right, and called it *nissology*. The islands were thus understood and promoted as the *focus* of scientific interest. The next steps aimed to validate the new discipline, and researchers looked to open-sea islands, which, with no mainland on the horizon, peaked out from the sea as almost untrammeled subjects of consideration. This choice also avoided problems that may have nothing to do with science but a lot to do with scientists. The world of academia is full of barriers that one would be advised not to impede. By moving far out to sea, nissologists eliminated potential criticism that they were trying to occupy ground someone else had already broken and who did not like anyone peering over the fence.

The very definitions of an island, particularly those that see island only as a part of the mainland surrounded by sea, can lead to the idea that open-sea islands are a welcome although easily exhaustible subject. If there is something that makes islands fundamentally different from the mainland, if it is at all possible to determine the island *differentia specifica*, then it is easy enough to discover and resolve all possible issues. The mainland definitions, however, point only to insularity, that is, weak links with the mainland, to the marginal position of islands in the networks established on the mainland, and to the difficulties that everyone encounters traveling from the mainland to an island and back, expecting it should be as easy as traveling overland. It is, however, pretty clear that insularity in this sense is simply a geographic circumstance that had initially determined development on the islands and made them so specific. Moreover, it is this specificity and not only insularity that awakes scientific curiosity and promotes islands as objects worthy of studying. New issues are constantly emerging and, with that, more research. Favorable responses are also accumulating; in its first thirty years, this brave new discipline has fought for its place under the academic sun, found itself included in university teaching programmes, and sparked the interest for new scientific journals, international societies and regular scientific meetings. Islands are no longer viewed merely as bodies of land surrounded by water, but as pieces of land that create intrinsic specifics by virtue of being encompassed. Insularity has conditioned many things: 1) the economy, which is difficult to establish, let alone develop, if resources are not used in a sustainable way;

2) producers and consumers who are universal in their sea-surrounded universe, and the universe of the sea around them; 3) the language that universality has enriched, and insularity preserved; 4) the endemic species that can hardly survive and certainly cannot thrive on the mainland; 5) the ecosystem that is exceptionally resistant in insular conditions but barely resistant at all to any *alochtoons*; 6) the symbiosis of nature and cultural heritage that is rarely found on the mainland; 7) the particular mentality of islanders; 8) the island identity that has been formed throughout the centuries and is different from that on the mainland; and, finally, 9) the self-sufficiency of islands built on the determination to survive on places surrounded by the sea. The social rules established by islanders have never been, and never could have been, significantly different from those on the mainland; however, due to the imperative of the sustainable use of resources, the islanders have had to align with the island ecosystem.

These specific features point to "islandness," the focus of nissology. It is a complex concept that describes the overall island existence in conditions of long-term insularity. It is complex and also quite dynamic, we might add, as it changes as we research it. The underlying reason for this is modern economic growth, which is too often rashly equated with development. This development has occurred on islands in the last fifty-odd years and dislocated what we term *islandness*. Rapid construction, comprehensive infrastructure, summer tourist invasions and autumn retreats after the mission of disrupting the environment has been accomplished, poor economic structures and the relentless exhaustion of island resources have affected the harmonious existence of the islands and islanders. Any nissologist, regardless of the novelty of their chosen discipline, will encounter the same sort of discouraging problems as natural scientists and especially ecologists: The subject they are dealing with is dissipating and decaying. The nissologist will also quickly notice that the developers' attacks on resources are always backed with some economic policy that fosters economic growth and comes in to preserve the environment when it is too late.

The space in which economic growth takes place is extending continuously and relentlessly to reach the farthest corners. The bearers of this extension, developers, tend to hug the coastline, but drawn by resources, they will inevitably launch out toward the islands and reach the near ones first. Upon arrival, they will enter a long-standing economic and social area, which the islanders may have wished to extend if the sea had not almost always halted them. The resulting conflict usually eats away at near island self-sufficiency. Mainland developers are not usually interested in island development or growth of island economies, but in how they can launch and conduct their own projects. They are often joined in such endeavors by a considerable number of islanders who are primarily interested in a more comfortable life. Island economies and communities, whether they want to or not, begin to specialise in and adapt to what is going on in the wider mainland-island economic space. A specialised island bears little resemblance to what it was in its past. Most of the necessities of life, which for centuries were produced on the island, are now sourced from the mainland, while those who purchase island products or services come from across the sea. Almost all newcomers and more and more native islanders are beginning to perceive insularity as the primary obstacle to growth (which they equate a priori with development), so poor connections with the mainland have become the main issue, with easy access to the mainland evolving to the primary development goal. One of the side effects of such development orientation is diminishing of interisland connections. Oriented toward the mainland in every aspect, near islands have less and less to exchange within the archipelago. Their poor interconnectivity is not seen as a problem, and it easily drops out of scope of the mainland policymakers. A specialised island in a once-homogenous archipelago thus becomes the subject of development policies that concentrate on satisfying the almost unison demand for better connections with the mainland. After being defined and labeled on the mainland, islands are subjected to deinsularising development policies that may or may not take islandness into account. Often, they do not: The main tasks are to eliminate the differences and obstacles seen by policymakers from the mainland as an "island handicap," reach the islands and link and integrate them with the mainland so that capital, goods, labour, information and whatever else is circulating in mainland networks can be safely delivered. It seems that defining an island as a part of the mainland surrounded by the sea is an augur of the future. Once achieved, the goals of a deinsularisation policy, which have (almost certainly) overlooked islandness, will confirm that islands have become a surrogate of the mainland. The policies that once addressed islands will be incorporated into the overall regional development scheme without the guarantee that it will be attuned to island specificities.

It is fortunate, however, that island development policies need not lose sight of islandness, no matter how relentlessly deinsularisation proceeds. The two are not mutually exclusive. Along with attempts to incorporate islands in the mainland, there are examples of respecting islandness and even signs of development policies founded on island specificities. However inevitable the disembarkation of developers from the mainland may seem, in the last few years we have witnessed the promotion of small islands to laboratories where *smart* technologies (sophisticated, yet robust, and can be relied upon in unfavorable conditions) can be developed, tested and applied. These can be upscaled and implemented on the mainland too and facilitate the development of smart towns much larger than small islands. In search of a nice acronym, policymakers made a good use of the words *sustainable*, *manageable*, *achievable*, *result-oriented* and *trackable*, turned into a label and then a flattering adjective awarded to both islands and towns. We are witnesses to the direct application of islandness: New technologies are being tested in challenging conditions that demand the island skill of using and preserving resources. Islands have turned into laboratories where innovations are developed and then applied on islands and the mainland. The compliment "smart" is indeed deserved.

Insularity, which can be overcome by building expensive infrastructure and introducing increasingly larger, faster transport connections, and islandness, which can be preserved with far fewer investments, each point in their way to two other island peculiarities. These are the size of the island and its distance from the mainland, neither of which are covered by the usual definition of an island and may seem trivial; however, size and vicinity reveal a great deal. The larger the island, the more economic reasons there are for including it in the mainland networks, which bolsters the call for deinsularisation from developers and development policymakers. Experience shows that islandness is at stake, and the policies that would preserve it are relegated to the background. On small islands, insularity is seen as a problem, but deinsularisation proceeds much slower. Small islands are too small to start anything alone and as a rule have too few resources to attract attention from the mainland. In this "game of thrones," where their role is minuscule, their trump card may be location, their position on the geopolitical map to which pretenders aspire, not for the welfare of the small islands, but to exert the closest possible surveillance of the seas or oceans. Islands far from the mainland, particularly tiny ones in the middle of the ocean, have been spared from massive waves of deinsularisation, but their islandness is threatened by something graver. They are exposed to the ill-omened danger on the islands themselves: Islanders who move away are rarely replaced with new islanders, and advanced aging on the islands is heralding imminent demographic demise. With the last islanders, the specificities and legacies apart perhaps from those wrought in stone, are slowly but surely fading away. Nissology, at least in terms of small and near, abandoned islands, is facing a reductio ad absurdum—islands without islandness.

How size and distance from the mainland determine island circumstances is best seen on islands that lie close to the mainland. They are the most numerous, generally speaking, and pose the most questions. For open-sea islands, the rule "sustainable or nothing" tends to apply. Their inhabitants have adhered to it and even institutionalised it. Near islands adhere to the same rule but with the ever-present opportunity of relying on the mainland. They are close enough to export items to the mainland, which is often a half-day of rowing away, so islanders have shown up on mainland markets as buyers and sellers or in the face of imminent shortages. They may also decide to make a living on the mainland while living on the island, taking their profits home with them on a weekly or even daily basis. It is close enough to allow them to go a step further—to move to the mainland to make a living and send money and provisions back to the island by the inertia of prudent island living. In time, they may ask themselves what a mainlander needs to do to become an islander and what an islander needs to do to establish roots on the mainland.

As a rule, the near islands have belonged to the states that possessed the mainland and occupied the coastline as widely as possible. The mainland economic networks reached them already in the age of sailing ships. The owners of small, near islands generally managed their estates from the mainland, respecting islandness only when it served their interests. Some near islands were so close that they ended up bridged, raising the issue of their islandness, since the imperative of sustainable use of scarce resources was no longer decisive.

The vicinity of the mainland makes near islands particularly interesting to not only nissologists, but also geographers, ecologists, regional scientists and development theoreticians. They should also be of interest, and sometimes are, to the development policymakers. The influence of the mainland weakens islandness and throws a mist over islands as such, indicating the need for further research. Such research may be more complicated than on the open-sea islands where nissology was born but should not lead to discouragement. The findings and knowledge gained far out in the sea are ample when moving closer to the mainland and focusing on islands that were perhaps not considered interesting enough to earlier researchers. The scientific baggage that nissologists bring with them from open-sea islands is enough, so there is no reason to fear that the mainland will succeed in distorting conclusions, concealing topics or reducing nissology to another existing discipline. There is no reason to fear that we will have to accept the legitimacy of the mainland approach to islands, abandon our near island focus and sail back out to sea.

Having established that an island far from the mainland is a worthy subject of research, we still need to respond to the question of whether an island that is close enough to reach by rowboat can develop the specificities that make it significantly different from the mainland. Has islandness been built and grown on such islands, and if so, has it been determined only by insularity? Has the mainland also left its stamp? To what extent is the island economy different from the mainland, and can the two be distinguished? Is the near island ecosystem self-sustaining, as on the Galapagos Islands, for example? Is the community on the near island developing skills of the sustainable use of scarce resources? Does the pattern of smart near islands have any prospect in the long term? There are many questions.

The task is neither insuperable nor easy. With near islands, we do not need to overcome a predisposition toward a view of utopias with abundant resources and a free lifestyle, as when dealing with open-sea islands. Near islands are just far enough away to be a place of refuge but too close to be considered utopian. There are other predispositions, however, and they may seem paradoxical. The islands are too close to be viewed as anything special. Many scientific disciplines observe them as extensions, sometimes even as anomalies, of the mainland. It is evident in development policies as well. Policymakers regularly set the proposition "island = mainland," implicitly or explicitly, and propose measures for deinsularisation, assuming that this will preserve islandness, or ignore the issue altogether. Since the notion of insularity necessarily implies the relation island–mainland, whereas the notion of island per se does not, considerations about near islands in their own terms may easily end up on a dead-end street. Focusing on near islands without taking into account the near mainland may deprive the analysis from important, perhaps decisive insights and lead to the conclusion that the entire problem will be solved by better transport connections. On a theoretical level the questions condense in one: What is the focus of nissology if the island is near? The island in its own terms or the immensely complex relations between the mainland and near islands and archipelagos? Indeed, we need both analytical minds and caution to stubbornly form new answers despite the likely rejection of the question.

These many questions lead directly to the Adriatic and the islands on its eastern side. Of the officially counted 1,246 islands, rocks and reefs that nestle along the coastline, only about fifty are inhabited. Geographically, they are almost in the centre of Europe and easily reached from the mainland, yet most of them have paradoxically remained on the margins of Mediterranean trends in almost every aspect. Despite this, or perhaps because of it, the relationships between the islands and the mainland, and the large and small islands, have always been and remain complex, each relationship being unique, just as the islands themselves are unique.

The Croatian islands have experienced virtually everything that can happen to islands anywhere: a tumultuous geological past; human presence dating back to the Paleolithic; the arrival of people who came to be islanders and those who wanted to subdue them; islanders who have survived using resources in the island space and those whose island existence has been supported by seafaring; the long-term mainland supremacy over islands; the occasional economic superiority of the larger islands over the mainland and the perfection of honed survival skills in the destitute island ecosystem and mainland political system that did not possess such skills but knew how to exploit them—slavery, feudalism, capitalism, socialism, transition, supplemented by uninvited landings that the islanders were not strong enough to repel.

There were exceptions, as the islanders autonomous enough to reach for the mainland tried to act in the same manner, setting up island–mainland civic communes under the Venetian doges. The fight for autonomy, which seemed easy enough on oceanic islands, worked in the Adriatic too, although the mainland states could hardly have been called peace-loving. In peaceful yet exciting periods of economic booms, the seafaring island entrepreneurs ventured forth on the world's oceans. In turbulent periods of changing powers on the mainland, the Adriatic islands often served as a refuge, with the sea as a defence wall. Even today, refugees perceive the sea as a defense feature and are keen to sail to an island to find shelter. When the COVID-19 pandemic reached worrying proportions in mid-March of 2020, many owners of second homes on the Croatian islands dashed across the sea to safety.

The Adriatic, long referred to as the Bay of Venice by doges and emperors, has always been a fractal of the Mediterranean containing the sea, diversified coast and an archipelago, a world for itself in which islanders created their own measures of nearness and farness. The islands, all near from a Mediterranean perspective, have always been approached as a set of subarchipelagos, some of them viewed as near, some quite near and some far away from the coast. The Adriatic perspective has also affected research; literature written by islanders about their own island, and no other, is plentiful. It tells of a quite heterogeneous archipelago where every island has its own story. The same goes for most of the mainland scientists that visit the islands. Social scientists, sociologists and anthropologists, in particular, tend to approach islands one by one, while economists tend not to approach them at all. The archipelago as a whole is dealt with far less, mostly by geographers and here and there by engineers searching for oil in the Adriatic. While the latter fortunately have not found anything so far, the former offered divisions that appear typical but do tell how diversified the archipelago is: near, channel, outer, northern, middle and southern islands, bridged islands and dredged islands, populated and unpopulated islands, islands versus rocks and reefs and the like. There is also a particular division imposed by such a heterogeneity: big, medium and small, the criterion being either size or population, or both. The Adriatic perspective is most evident here: The biggest island appears small in the Mediterranean, let alone the world perspective.

Regardless of the perspective, the Croatian archipelago deserves more than a single book. We will thus skirt around many islands and entire subarchipelagoes without justification, not being able to do more; more space would be needed to write about everything worthy of record. The historical luxuriance, rich heritage and island culture, which are slowly disappearing between the cypresses in island graveyards, along with the good and unfortunate things brought from the mainland by modern immigrants, new islanders, will be briefly touched on in this book. After getting a view of the world and, in particular, the European near islands, we will learn about the geography of the Croatian islands and how islanders were forced to accept migrations as a way of life. We will then move our focus to specific islands and learn about islanders' identity; self-divided Hvar; Silba, the island of seafarers; Olib, the island that moved to America; the Zadar urban archipelago; the fishing islets in the middle of the Adriatic, where island space consists of a couple rocks and a lot of sea around them; and Susak, the exception to them all. It is an archipelago of chapters about an archipelago where we will come across uninvited landings, omnipresent fateful emigration and the influence of the mainland, where policies for the use of island resources have been made for centuries. This will lead us to the underlying problem of the near islands; despite all the research, nissological findings, laws and policies, they are still too close to be recognised as different from the mainland. Another problem is the communication launched across the sea by the islanders themselves—plenty with the mainland and little between the islands. The reader will get to know an apparently homogeneous archipelago where the islands seem quite alike, with similar resources and production, where none can offer the others anything they do not already have. The mainland has always been a more promising buyer and seller, so the islanders have always turned more readily to it than to one another. This made them so heterogeneous, and perhaps this is why so few islands and islanders could spin such diversity.

If we set off from a Croatian island harbour for the mainland, we will see that the mainland, moreover, the river experience, has in no small extent defined the current notion of near islands. Many things on them have been noted as different—many things except the islands themselves. With rare exceptions, they are still the locus of research, which this book aims to overcome. In the pages that follow, the Croatian islands in their own right will be the focus at all times.

REFERENCES

Baldacchino, Godfrey. 2013. *The Political Economy of Divided Islands: Unified Geographies, Multiple Polities*. London: Palgrave Macmillan.

Baldacchino, Godfrey. 2006. "Editorial: Islands, Island Studies, Island Studies Journal." *Island Studies Journal* 1, no. 1: 3–18.

Baldacchino, Godfrey, ed. 2007. *Bridging Islands: The Impact of Fixed Links*. Charlottetown, Canada: Acorn Press.

Deleuze, Gilles. 2004. *Desert Islands and Other Texts, 1953–1974*. Los Angeles, CA: Semiotext(e).

Grydehøj, Adam. 2018. "Islands as Legible Geographies: Perceiving the Islandness of Kalaallit Nunaat (Greenland)." *Journal of Marine and Island Cultures* 7, no. 1: 1–12.

Hay, Pete. 2006. "A Phenomenology of Islands." *Island Studies Journal* 1, no. 1: 19–42.

Hess, A. L. 1986. "Overview: Sustainable Development and Environmental Management of Small Islands." In *Sustainable Development and Environmental Management of Small Islands*, Man and Biosphere Series, vol. 5, edited by W. Beller, P. d'Ayala, and P. Hein. Paris: UNESCO.

Lewis, Martin W., and Kären E. Wigen. 1997. *The Myth of Continents: A Critique of Metageography*. Berkeley: University of California Press.

Matvejević, Predrag. 1987. *Mediteranski brevijar*. Zagreb: Grafički zavod Hrvatske.

McCall, Grant. 1994. "Nissology: The Study of Islands." *Journal of the Pacific Society* 17, no. 2–3: 1–14.

McMahon, Elizabeth. 2010. "Australia, the Island Continent: How Contradictory Geography Shapes the National Imaginary." *Space and Culture* 13, no. 2: 178–87.

Morris, F. Ellen. 2010. "Insularity and Island Identity in the Oases Bordering Egypt's Great Sand Sea." In *Thebes and Beyond: Studies in Honour of Kent R. Weeks*, edited by Zahi Hawass and Salima Ikram, 129–44. Cairo: Supreme Council of Antiquities Press.

Ronström, Owe. 2009. "Island Words, Island Worlds: The Origins and Meanings of Words for 'Islands' in North-West Europe." *Island Studies Journal* 4, no. 2: 163–82.

Royle, Stephen Arthur. 2007. "Island Definitions and Typologies." In *A World of Islands: An Island Studies Reader*, edited by Godfrey Baldacchino, 1–29. Charlottetown, Canada, and Luqa, Malta: Institute of Island Studies, University of Prince Edward Island and Agenda Academic.

Stratford, Elaine. 2003. "Editorial: Flows and Boundaries: Small Island Discourses and Challenges of Sustainability, Community and Local Environments." *Local Environment* 8, no. 5: 495–99.

PART I

Chapter One

On Near Islands

Stephen A. Royle

There are relatively few islands in the world that are remote from any mainland coast. This chapter discusses the more numerous cases of "near islands" whose proximity to a mainland coast places them under the influence of that mainland to some degree. A global perspective is taken with examples from throughout the world, if with some focus on my home island of Ireland—a near island to Great Britain—while at a different scale, Ireland's many small offshore islands are near islands to Ireland itself. Croatia and its islands are not featured. The generalities of "near islands" will be applied subsequently to the Croatian case, when the scope will be narrowed to the Adriatic Sea.

The complexity related to the broad and widespread use of the concept of "island" is matched by the way in which actual bodies of land surrounded by water have been categorised. One obvious differentiation relates to size: Large islands are seen as being distinct from small islands, with the generally accepted implication that the smaller bodies of land are most impacted by the restrictions (or opportunities) associated with islandness; however, there are varied views as to what might be the threshold size between large and small islands, and whether area or population total should be the factor upon which the threshold should be based. Greenland, at more than 2 million km^2, is not small in area; indeed, it is generally regarded as the world's largest island (assigning Australia to a position as the smallest continent), but with only about 56,000 inhabitants, with regard to population, Greenland is small (Karlson 2009).

Other differentiators between islands would be geology, political status and economic type. A volcanic island presents different challenges and opportunities compared to, say, a coral atoll. An island that has full political autonomy has a different range of economic possibilities to one that has a dependent status. An island that focuses on the production of, say, sugar presents differences to an island engaged in mass tourism or another that is a military base.

These categorisations are filtered through location. An island within an archipelago relates to being a land mass surrounded by water in a different fashion to a single island either off the coast or set apart in the middle of the ocean. In an archipelago, facilities can be spread throughout different islands. Thus, in the Maldives in the Indian Ocean, the international airport is on, indeed it is co-terminus with, Hulhulé Island with ferry connections to the nearby capital island of Malé, which, with a population density of about 23,000 per km^2, certainly has no room for an airport. Other Maldivian islands are dedicated to such specific uses as resorts. Compare this situation with a lone island like St Helena in the South Atlantic Ocean, where necessary functions and facilities have to be located within its single landmass of 122 km^2, 2,000 km from the African continent. To take the case of airports again, that on St Helena opened to commercial flights only in 2017, despite consideration first having been given to building an airport in 1943, with the project being revived in the 1960s. One issue was finding the political will to develop the project on this British Overseas Territory. Another major difficulty was identifying where an airport might be located on this mountainous island. Furthermore, a great deal of money had to be raised to overcome the logistical difficulties of not only actually constructing an airport amid St Helena's rugged topography, but also dealing with the island's remoteness and the consequent long supply chains.

NEAR ISLANDS

Another set of island polarities would be distance from a continental (or large island) shore. Remote, open-sea islands, characterised by Patrick Nunn as "oceanic islands" (Nunn 1994), are opposed to the more numerous "near" islands, those only a relatively short distance from a continental (or large island) coast. Literature on deep-sea islands often emphasises remoteness and isolation. *Island on the Edge of the World* is about St Kilda, the most distant lands of the Scottish Western Isles (Outer Hebrides), 64 km northwest of North Uist in the main Western Isles chain (Maclean 1977). The population of Hirta, the only inhabited island of the St Kilda group, was taken off in 1930, life there having become too hard and difficult to be sustained in the modern world (Steel 1975). *Island at the End of the World* is about Rapanui (Easter Island) (Fischer 2005), one of the most remote of all islands, more than 3,500 km from Chile. Paradoxically, before European discovery in 1722, Rapanui's Polynesian inhabitants conceived of their island as the extruding navel of an otherwise entirely watery domain, being thus the centre rather than the "end of the world."

The position of near islands, located close to their adjacent mainland, brings about a different set of circumstances compared to the remote islands. St Helena and Manhattan are both surrounded by water, but that is the only feature they have in common. As stated, St Helena only recently transitioned from sea to air access, while Manhattan is connected to mainland New York and New Jersey by more than 20 bridges and tunnels.

Nearness, of course, is a relative term dependent upon prevailing circumstances that can change throughout time especially regarding technological development and infrastructural facilities. Take as an example, Yakushima, which is a mountainous island of the Osumi group in Kagoshima Prefecture, Japan. It lies 61 km from the "mainland" of Kyushu Island (in Japan the four large, interlinked islands of Kyushu, Shikoku, Honshu and Hokkaido are viewed as the mainland). In past times, this distance was sufficiently isolating for Yakushima to have developed a distinctive fauna with some notable speciation, not least in the development of a subspecies of macaque, the Yakushima macaque (*Macaca fuscata yakui*), and a variety of sika deer. In addition, the island has about 50 species of endemic flowers, and within its extensive surviving primary forest are both endemic and notably ancient trees. One *Criptomeria* (Japanese cedar), some thousand years old, has its own name—Jomon Sugi—and an individual Wikipedia entry that traces Yakushima's now-significant tourist industry to its discovery in 1968. Since then, environmental conservation has been significant on its island, including the curtailment of forestry and a newly developed regard for the macaque and deer populations, which are now protected. Such protections are necessary, as the environment of Yakushima—the product of its former isolation—is now stressed by its modern relative closeness to mainland Japan, given that the island's splendid landscape, flora and fauna attract thousands of tourists each year. It is not that the island has moved nearer to the mainland, of course; it is just that now the 61 km stretch of sea is much less of a barrier, and it is crossed many times a day by hydrofoils, car ferries and aeroplane services from the city of Kagoshima and other places. Yakushima has become a near island, heavily influenced by its mainland.

Other islands that have become "nearer" include Nantucket and Martha's Vineyard off the coast of Massachusetts in the eastern United States. These two islands are now upmarket resorts focussing on serving a wealthy clientele who travel there easily by boat or air in the thousands, especially in the summer high season; however, in the early decades of European settlement it was not the islands' accessibility that stimulated their economies, rather it was their isolation, for both islands were significant bases for whaling. Herman

Melville's classic 1851 whaling novel *Moby Dick* is set partly on Nantucket, and some of the chief (human) characters are from the island (Melville 1851).

A further example of the relativity of nearness comes from a study of islanders' identity in two islands offshore the large island of Ireland as the tyranny of remoteness has eased. In the past "identity was structured by the shared hardships and enforced self-sufficiency associated with residence in communities that were relatively remote from the mainland." Matters changed with modernisation, including improvements in transportation, and the study showed that "contemporary islander identity reflected the increasing diversity of small-island populations. Over time the relative remoteness of island communities has diminished and new populations have moved in" (Burholt, Scharf and Walsh 2013, 11).

FIXED LINKS

Nearness certainly brings opportunity to islands, the main one being ease of access, especially if the near island can be tagged to the mainland by fixed links. Physical linkage impacts islands, as functionally a bridge, tunnel or causeway turns an island into a peninsula or an archipelago into a single island. Bermuda would be regarded as a single island and certainly functions as one, but in reality, it is an archipelago, with many of its islands, including the major ones, linked together; however, it can be debated as to whether the provision of a fixed link violates the identity of the (former?) islanders. Did the approximately 150,000 residents of Prince Edward Island in Atlantic Canada lose their distinctiveness as islanders in May 1997, with the opening of the 12.9 km Confederation Bridge? This stretches across the Abegweit Passage of the Northumberland Strait and connects Borden-Carleton in the western part of the island to Cape Jourimain in New Brunswick.

A plebiscite had been held on the island in 1988, which found 59.5% of respondents in favour of a fixed link, so there was popular support for the project, although a substantial minority always remained adamantly opposed. A representative of an antibridge pressure group, Friends of the Island, said, "God separated us as an island. If He had wanted to hook us up, He would have done it at the time." In a report it was noted that the island Friends envisioned an influx of drugs and prostitutes, and the end of the "island way of life" (Friscolanti 2014). Certainly, some common features of island life disappeared, for example, the principal year-round ferry service to the island, which used to run from Borden-Carleton to New Brunswick. Fisheries, particularly for lobster, were affected as had been feared, but the bridge is now an accepted part of life on Prince Edward Island: "Once it was there you might as well use it" (Friscolanti 2014). Tourism has benefitted from the increased—and eased—accessibility to the mainland, particularly the ending of queuing for the car ferry, which was a considerable disincentive to travelers. The "fact of difference" an island enjoys because of its isolation, associated with mystery, even romance—"paradise island," "treasure island," "island of dreams"—has been seen as important in both the identity of islanders and the appeal of islands, and it has been suggested that islands with fixed links lose this (Baum 1997). Nonetheless, Prince Edward Island residents retain a distinctive identity and culture ascribed to them by their "Island," always written in the region with an initial capital letter as if to stress its significance.

Another island with a fixed link in the Canadian Atlantic Provinces is Cape Breton Island, which has been connected by the 1.3 km Canso Causeway to Cape Porcupine on mainland Nova Scotia since 1955. Cape Breton has a distinct heritage and identity. It was populated mainly by Scots forced from their native land by the Highland Clearances when tenant farmers were evicted by estate owners in the 18th and early 19th centuries. Scots Gaelic was a first language on the island until fairly recently, and the Scottish heritage there is both strong and cherished. In addition, Cape Breton has had a distinct economic history focused on coal mining and steelmaking. These industries have now closed but have left their marks on both the island's landscape and the islanders' identity and psyche—as revealed tellingly in short stories by island-based author Alistair MacLeod (1936–2014), as well as in his more substantial work, *No Great Mischief* (recognised as Atlantic Canada's greatest novel) (MacLeod 1999, 2000). Despite the fixed link and its nearness, Cape Breton remains a place apart from mainland Nova Scotia. No wonder then that when the Canso Causeway opened, an elderly islander was recorded as thanking God for "having at last made Canada a part of Cape Breton" (Lotz and Lotz 1974, 14). Near islands, upon being connected by fixed links, thus do not necessarily lose their insular identity.

In many cases, fixed links have positive impacts on islands in that improved accessibility has been seen to lessen emigration. Fixed links position islanders closer in terms of time and convenience to mainland services, while island producers can more readily access their markets. The islands themselves are made easier to reach from the mainland, which encourages tourism and integration. Godfrey Baldacchino has edited a book about island fixed links, and Table 1.1 is adapted from it and considers the impact of fixed links on the near island fringe off the north and west coasts of the large island of Ireland (Royle 2007). It shows that since the census of 1841, which was before the demographic catastrophe of the Irish famine later that decade, the population living on the Irish islands and the number of islands inhabited have both fallen dramatically; however, the fall has been more on unlinked islands compared with those having a fixed link, and in recent years about two-thirds of the Irish near island population has lived on islands that are linked.

Table 1.1. Irish Islands, 1841–2016

	1841	1926	2016
Number of inhabited islands	180	112	76
Total island population	35,828	21,022	8,756
Number of linked islands	13	20	24
Population on linked islands	5,037	8,622	5,964
% living on linked islands	14.1	39.2	68

Source: Royle (2007).

NEAR ISLAND–MAINLAND RELATIONSHIPS

A near island is inevitably influenced by its mainland, a situation that is not always positive. Proximity can engender dependency, with near islands becoming subject to dominant mainland influences, leading, in the limiting case, to such islands being or becoming mere extensions of the mainland. It is apposite here to note that John Donne, having asserted that "no man is an island," went on to write, "Every man is a piece of the continent, a part of the main[land]." This might happen literally to a near island if land reclamation takes it into the mainland. One example is Macau. This territory, on the western side of the Pearl River estuary in southern China, was leased to Portugal in 1557, and granted in perpetuity in 1887, although it was returned to China in 1999. Macau originally comprised three islands, Macau itself and, farther from the coast, Coloane and Taipa. Macau island has been a peninsula since the 17th century, when land reclamation involving a natural sandbar connected it to the mainland. Later, bridges were built out to Taipa, the closer of the other islands, with a causeway onward to Coloane; however, since 2005, the area between the two once-separate offshore islands has been reclaimed, being called Cotai, a combination word with elements from both former separate islands. What was once a geography comprising three islands off mainland China has been transformed; Macau now really is a "piece of the continent."

IRELAND AS A NEAR ISLAND

Mainland influences on near islands are not always physical. Power and scale differentials between the island and the mainland might see the island dominated, exploited in some way. One relevant, large-scale example can be seen in the long-standing, disputatious relationship between the major island of Great Britain and the smaller, near island of Ireland to its west, at its closest point only 21 km away. British domination of at least parts of Ireland goes back centuries. There was an invasion from England in 1169, and in the Middle Ages a large part of eastern Ireland was under direct British control with the imposition of external cultures and customs, including the English language and the settling of Ireland with people from England and later Scotland, a process called the plantation (Montaño 2011). This reached its peak in the 17th century, when in the province of Ulster in the northeast, land confiscated from a local lord who had favoured Spain in the Anglo–Spanish War was reissued to settlers from Great Britain. This process, the Plantation of Ulster, saw tens of thousands of Protestants move into what had been a largely Catholic country, reinforcing ethnic divisions within Ireland. Quotidian sectarianism occurred whenever and wherever the competing groups of Catholics and Protestants faced their fellow Christian, physically indistinguishable, Other. In addition, there was colonialism, land grabs, forced population movements and ethnic cleansing. There were rebellions within Ireland in the mid- and late 17th century, and after another rising in 1798, the unequal islands and their offshore isles were brought together as the United Kingdom of Great Britain and Ireland under the 1801 Act of Union.

In this United Kingdom, the near island of Ireland was certainly the junior partner, although the Irish people were of great utility to Britain as a labour force. They built the canals and the railways, worked in the factories, served disproportionately in the armed forces and migrated seasonally to gather the harvest. Relations were not helped by the failure of the mainstay of Irish peasant life, the potato, in the Great Famine of 1844–1849. The British disliked the necessity of having to provide support, while the Irish resented the clumsy and ill-run attempts at relief from Britain. Pressure mounted in favour of Irish independence, encapsulated under the label of Home Rule, but this had not been granted by the outbreak of the First World War in 1914. During this conflict came an armed insurrection in Dublin in 1916, known as the Easter Rising. This was put down harshly, which made the British even less popular, and after the war, the Irish Republican Army began a campaign of violence, the War of Independence. The United Kingdom granted Home Rule to most of Ireland under the Government of Ireland Act of 1920, which founded what became the Republic of Ireland. The island of Ireland was partitioned; six of the nine counties of the province of Ulster, which had a Protestant majority, remained within the reduced and renamed United Kingdom of Great Britain and Northern Ireland. The sacrifices made by Ulster Protestant soldiers in the First World War, especially at the Battle of the Somme in 1916, had made it impossible politically to force their home area into a united Ireland. There was also a realisation that there would be massive, armed resistance if much of Ulster was to be expelled from the United Kingdom. Partition saw tens of thousands of people trapped on what they regarded as the "wrong" side of the border, including about one-third of the population of the city of Belfast who were Catholics. There was civil disturbance within Northern Ireland as a result with violent IRA campaigns seeking Irish unity in the 1920s, 1930s and 1950s, and then came the "Troubles," a mild name for almost 30 years of violence beginning in the late 1960s.

To conclude, Ireland's history has long been bound up with that of its much more powerful eastern neighbour; it was a "near island." But after both the United Kingdom and the Republic of Ireland joined what became the European Union (EU) in 1973, the Republic of Ireland fervently embraced a European identity. The United Kingdom, by contrast, was never an enthusiastic member of the EU, and in 2016, a referendum saw the country opt to leave, albeit by a small majority, a process known as Brexit (BRitish EXIT). This meant that the border within Ireland became an international land boundary between the EU and an outside state. In negotiations about how to deal with this, the Republic of Ireland, for so long the majority part of the dependent and relatively powerless near island of Ireland, but now a committed member of a greater, pan-European polity, stood up to its former ruler. Ireland is a near island no more.

CAPITAL ISLANDS

This situation of unequal power and influence can also be seen to develop within island groups, including multi-island states where the capital island has become dominant, with other islands displaying the dependent relationship typical of near islands. There can even be the situation where the dominant island becomes synonymous with the group as a whole; Tahiti would be a more widely recognised place name than French Polynesia, for example. The Marshall Islands and Kiribati are Pacific island nations largely composed of atolls, where the capital islands of Majuro and South Tarawa, respectively, have the most developed economies and offer employment opportunities and urban facilities not widely available elsewhere in their nations. These capital islands have become focal points for immigration from the other islands, and their populations have grown to an alarming extent given the delicate environment and limited resource base typical of coral atolls. The dependent off-islands suffer from population stagnation or decline, in all a lose-lose situation. As long ago as 1969, it was recognised for Kiribati that "urbanisation is seen as deruralisation. People leave social and physical resources in the rural areas [the off-islands] (which then decay) to go to nonexistent social and overstrained physical resources in the town" (Hughes 1969, 84). Table 1.2 shows the continued and increasing dominance of the capital island of South Tarawa within Kiribati.

Table 1.2. Kiribati Island Populations, 2010 and 2015

Island	*2010 Population*	*% of Total*	*2015 Population*	*% of Total*
South Tarawa	50,182	48.7	56,388	51.2
Other islands	52,876	51.3	53,748	48.8
Total	103,058		110,136	

Source: Kiribati National Statistics Office (2016).

RESENTMENT

Just occasionally, the dependency relationship between near islands and the mainland operates the other way, with the nearby mainland being dependent. One Canadian example comes from the southern coast of Labrador and the easternmost part of Quebec, where an isolated series of mainland villages on the north shore of the Strait of Belle Isle from Blanc Sablon in Quebec east to Red Bay, Labrador, depend for their physical contact with the outside world on the ferry service from Blanc Sablon to St Barbe on the island of Newfoundland. Similarly, in the Highlands of Scotland one coastal mainland village, Inverie, has no link to the national road network and, like such nearby islands as Eigg and Rum, depends on a ferry service to the mainland port of Mallaig for outside access. Thus, part of the mainland operates as a near island.

Hosting a ferry port is an economic benefit to a mainland settlement; however, given that, in general, ferry services cross as small a stretch of water as possible, the mainland terminal might be located in a place chosen only in regard to its proximity to the island with little else to commend it. This can lead to resentment, as depicted in a book by Deborah Tall (1996), who detailed such a relationship in regard to the Irish island of Inishbofin, concealed imperfectly in her account as *The Island of the White Cow* (which is what *inishbofin* means in Irish):

> Clochan [on the mainland, in reality Cleggan] is a grocery, a butcher, a pub, a pier, and a small parking lot for cars that have been abandoned in favour of the island. Clochan is a place indeed, largely in existence for the sake of the island, and its residents' faces are bitterly etched with that sad fact. The people of Clochan despise the tourists who are in their village only on sufferance, though

tourists provide the only money made in the place: hours of drinking in wait of the island boat. (6–7)

Resentment of an island by the adjacent mainland can also be found elsewhere. One historical example comes from Western Canada. Off the continental coast of British Columbia lies Vancouver Island. From the late 1840s, this island was administered by the Hudson's Bay Company under a mandate from the British crown. The company had a requirement to stimulate European settlement, in addition to carrying out their customary fur trading and other business with the island's and surrounding regions' indigenous peoples. The company struggled to fulfil this dual role, especially as attempts to open up coal mining ventures did not meet with success at that time. Nonetheless, their settlement of Fort Victoria was developed in the south of the island and served as the capital of Vancouver Island Colony. The adjacent mainland, then called New Caledonia, although nominally part of British North America, was largely empty of Europeans. Then, in the late 1850s, gold was discovered on the mainland along the Fraser and Thompson rivers. The subsequent gold rush led to considerable immigration of Europeans, initially through Vancouver Island, to the considerable economic benefit of Victoria. Given that the mainland now had to be properly administered, it was declared to be a new colony, British Columbia (BC), in 1858.

For a few years this was a separate polity to Vancouver Island Colony, but in 1866, for administrative efficiency, the two colonies were to be merged under the mainland colony's name. One American newspaper report regretted this, if with a sneer at the near island, within a perceptive recognition of mainland–island relationships. Union would "sweep away the constitution, the free port and everything that attracted capital and intelligence to that out-of-the-way island and to give the people over, bound hand and foot, to the domination of the Governor of BC" (*San Francisco Bulletin* 1866). The two colonial governors did not agree as to the best location for the capital city of the unified colonies. The choice was between Victoria—which had long outgrown the extent of and the name of Fort Victoria—on Vancouver Island and the newly formed settlement of New Westminster (forerunner to Vancouver) on the British Columbia mainland, which had been the capital of the separate British Columbia colony since 1859. Arguments in favour of the island settlement focused upon its greater level of development at that time, as well as its more convenient location for trade, including links with the outside world. Pro-mainland voices spoke of the need to favour the development of the potentially larger colony and not to disincentivise investors who had seen its potential. Moreover, there was a perceived benefit for future national and international relationships of having a mainland capital rather than, as the governor of British Columbia put it, the mainland being seen as the "dependency of an outlying island." No doubt to the dismay of the overweening mainlander, the island city of Victoria was chosen as (and remains) the capital of British Columbia, although the city of Vancouver, of which New Westminster is a suburb, is more than seven times its size (Royle 2011).

Relations between an island and a nearby mainland might be worse than mere resentment. The Channel Islands—Jersey, Guernsey, Alderney and several smaller islands—have been possessions of the British crown since the 13th century despite lying much closer to France than Southern England. England (later the United Kingdom) has spent a great deal of its history at war with France, and thus what geographically would seem to be near islands of France have had to be heavily defended against their neighbour. The Channel Islands landscape bears witness to this in a considerable amount and range of military architecture. The most recent fortifications are German, for in the Second World War the British had to take the painful decision to abandon the islands, as it was deemed impossible to protect them against the German forces, which had occupied France.

Another example where proximity is not the controlling factor in island–mainland relationships comes from North America. Off the southern coast of Newfoundland, 25 km from the Burin Peninsula, lies the small archipelago of Saint Pierre et Miquelon. France used to possess extensive territories in what is now Canada but relinquished them following defeat in the Seven Years' War, which was concluded by the 1763 Treaty of Paris. Only Saint Pierre et Miquelon was retained as a base from which to access fishing grounds, although the islands were subsequently fought over by Britain and France a number of times. France finally secured the by then depopulated islands in 1816, and settled them once more largely from Brittany and Normandy. Today the islands are an "overseas collectivity" of France, and the 6,000 inhabitants display in a determined manner French culture and customs just that short distance from the coast of Canada.

Returning to Ireland but lessening the scale, the general situation of the relationship between the "mainland" of Ireland and its fringe of small coastal islands might be considered. Many of these were depopulated, and all have lost population since the Irish famine of the 1840s. By the 1980s, the remaining populated near islands of the Republic of Ireland were, in political and administrative terms, mere appendages of their mainland local authorities, the county councils. Then, in the mid-1980s, representatives from different islands began to band together, to scale up their limited individual influences, and they set up a pressure group, *Comhdháil na nOileán* (the Federation of Islands), to seek assistance at the national level for common areas of support all the islands needed. Progress was not always smooth or steady, but a level of success can be traced back to those early efforts insofar as the true Irish islands—those without a fixed link—now enjoy dedicated ministerial oversight. The islands have a place within the Department of Culture, Heritage and the Gaeltacht, a central objective of which is to "ensure that sustainable vibrant communities continue to live on the islands" (Department of Culture, Heritage and the Gaeltacht). Support includes subsidised access services. The EU also supports the Irish islands.

Another set of islands that enjoy support are those off Taiwan, given the operation of its Offshore Islands Development Act (Royle and Tsai 2008). One of the most interesting cases is that of an island called Orchid Island, Lanyu or Pongso no Tao. It is the homeland of the Tao ethnic group, which is distinct from the mainly Han Chinese population of Taiwan itself. The Tao make up about 80% of the island's population of 5,000. Their principal economic activity is the exploitation of flying fish, and this fishery is protected by the Taiwanese government in that mainland boats are not allowed to catch. The fish are still taken by traditional means by island fishers using small boats they have constructed themselves. Island businesses are protected from outside competition, and the islanders receive free electricity. On the face of it, then, the situation on this near island is positive; however, there is a darker side, too, that serves to illustrate the dismissive relationship often seen between mainlands and their near islands. In 1982, without consultation with any islanders, Taiwan built a nuclear waste storage facility on the southern tip of Orchid Island. The islanders have protested this for years, but the waste remains.

CONCLUSION

This chapter has shown that islands have been categorised by such things as size, geology, political status and economic type, filtered through location, from archipelagos to single islands and from deep-sea islands to near islands, adjacent to continental coasts or larger, more dominant islands. Near islands are in a beneficial position in some regards, for instance, ease of access, especially if they can be connected by a fixed link; however, near islands, because of inequitable power relations with their adjacent territories, have been subject to domination from outside, in some cases causing resentment or even violence. I suggested elsewhere that islands everywhere are subject to a similar range of challenges because of the nature of islandness (Royle 2001). This was exemplified when in early 2019, I was invited to participate in a BBC World Service radio piece about the small Irish islands. News travels fast these days, and the sea between the mainland and islands is no more a barrier. The recording of the interview reached Croatia, and the comment bounced back to Ireland the same day. A Croatian friend responded, "Your interview and the whole broadcast sounded so familiar, as if it was about the Croatian islands."

REFERENCES

Baum, Tom. 1997. "The Fascination of Islands: A Tourist Perspective." In *Island Tourism: Trends and Prospects*, edited by D. G. Lockhart and D. Drakakis-Smith, 21–36. London: Pinter.

Burholt, Vanessa, Thomas Scharf, and Kieran Walsh. 2013. "Imagery and Imaginary of Islander Identity: Older People and Migration in Irish Small-Island Communities." *Journal of Rural Studies* 31: 1–12.

Fischer, Stephen Roger. 2005. *Island at the End of the World: The Turbulent History of Easter Island*. London: Reaktion.

Friscolanti, Michael. 2014. "God Separated Us as an Island." *Macleans*, 28 June 2014, https://www.macleans.ca/news/canada/p-e-i-and-the-confederation-bridge-debate/ (accessed 14 February 2019).

Hughes, A. V. 1969. "Unpublished Minute." Cited in E. E. Bailey. 1983. *Report on the* [Kiribati] *1978 Census of Population and Housing, Volume III: Analytical Report.* Tarawa: Ministry of Finance.

"Islands." *Department of Culture, Heritage and the Gaeltacht*, https://www.chg.gov.ie/gaeltacht/islands/ (accessed 14 February 2019).

Karlson, Agneta. 2009. "Sub-National Island Jurisdictions as Configurations of Jurisdictional Powers and Economic Capacity: Nordic Experiences from Åland, Faroes and Greenland." *Island Studies Journal* 4, no. 2: 139–62.

Kiribati National Statistics Office. 2016. *2015 Population and Housing Census*. Tarawa: Ministry of Finance.

Lotz, P., and J. Lotz. 1974. *Cape Breton Island*. Newton Abbot: David and Charles.

Maclean, Charles. 1977. *Island on the Edge of the World: The Story of St Kilda*. Edinburgh: Cannongate.

MacLeod, Alistair. 2000. *Island*. Toronto: McClelland and Stewart.

MacLeod, Alistair. 1999. *No Great Mischief*. Toronto: McClelland and Stewart.

Melville, Herman. 1851. *Moby Dick; or, the Whale*. New York: Harper Brothers.

Montaño, John. 2011. *The Roots of English Colonialism in Ireland*. Cambridge, UK: Cambridge University Press.

Nunn, Patrick. 1994. *Oceanic Islands*. Oxford, UK: John Wiley.

Royle, Stephen A. 2014. *Islands*. London: Reaktion.

Royle, Stephen A. 2011. *Company, Crown and Colony: The Hudson's Bay Company and Territorial Endeavour in Western Canada*. London: I. B. Tauris.

Royle, Stephen A. 2007. "Islands Off the Irish Coast and the 'Bridging Effect.'" In *Bridging Islands: The Impact of "Fixed Links,"* edited by Godfrey Baldacchino, 203–18. Charlottetown: Acorn Press.

Royle, Stephen A. 2001. *A Geography of Islands: Small Island Insularity*. Routledge Studies in Human Geography. London and New York: Routledge.
Royle, Stephen A. 1999. "Bridging the Gap: Prince Edward Island and the Confederation Bridge." *British Journal of Canadian Studies* 14, no. 2: 242–54.
Royle, Stephen A., and Huei-Min Tsai. 2008. "Cultural Heritage and Development Strategies of the Offshore Islands of Ireland and Taiwan." *Insula, International Journal of Island Affairs* 17, no. 1: 47–56.
San Francisco Bulletin, 20 August 1866.
Steel, Tom. 1975. *The Life and Death of St Kilda*. London: HarperCollins.
Tall, Deborah. 1986. *The Island of the White Cow*. London: André Deutsch.

Chapter Two

Near Islands of Europe

Sean Turner

The island landscapes of Europe hold some of the most remarkable environments and biodiversity values, preserve and maintain some of the oldest aspects of civilisations come and gone, and help spin a narrative of how the continent emerged and grew. Godfrey Baldacchino (2013, 17) refers to the "European project" and calls islands the sites where some of its most profound cultural and political challenges are met. Correspondingly, nearly every coastal European state has an island or islands that constitutively figure into their history, although these islands are typically the reasoning for some territorial dispute past or present and/or far removed from the mainland. It is near islands, those closest to state borders and that most often form maritime borders, that are often overlooked in both policy and academic spheres. That is not to say that near islands do not come into academic focus. They do, as they contain many prevailing state or world customs and are certain nodes of significance during some state territorial expansion; however, these are the tales of islands that added value to a state, the value being land, resources or strategic position where there previously was none. For instance, there is the example of Rockall, an uninhabitable islet in the midst of the Atlantic measuring a mere 784.3 m^2 but valuable enough to set off a mad dash for possession due to possible oil resources and, arguably more important, the additional maritime territory of 12 nautical miles afforded to island spaces.

While the previous example seems anecdotal, it represents how pursuance of distant land areas, no matter their size, can spur action if it adds value to the state. Contrarily, near islands are already incorporated into overall state value, and so their nature appears to be assumed in the same manner of every other mainland area if available research is any guide. As such, near islands are desperately underresearched and thus chronically misunderstood, especially if they do not substantially add to an overall land area. Nevertheless, even those that do can still fall victim to this non-value-added trap. For example, Denmark owes a great deal of its overall land area to islands, and the island of Zealand houses its capital; however, a cursory search in any major database will yield results for a Zealand a half-world away, and searches for "Danish islands" will mainly deliver research for the Faroe Islands. It is this ill-conceived presumption that there is little to be told from small distances that seems to guide research interests and possibly the subsequent mainland responses that manifest from it.

What is of interest here is the relationship of near islands to individual states and not the island state to the continent. This distinction is significant to make clear what needs to be established, which is that the near islands of a mainland state are of unique distinction. A near island, while possessing all the trappings associated with boundedness, physical isolation, and insularity, typically does not have the autonomy to fully chart its course. European states each utilise individual paths for island management, from consistency with any other land area incorporated into the overall state governing scheme or unique management strategies built to acknowledge and address the particularities of island spaces. Thus, while mentioning a near island, the mainland must at all times also be equally present, as it is the foundation of this concept of nearness, answering the question, "Near to what?"

This notion of nearness and its impacts bring about a scarcely touched upon idea for near islands: two terms. Since the inception of the field of nissology—the study of islands and islandness—discussing and studying islands on their own terms has long been the guiding principle, a concept introduced to confront the common misunderstandings of islands and lack of organised knowledge about them (Hall 1994). As such, cognisance of islands since the outset has come from observations gathered far out at sea. But as Baldacchino (2004) notes, islands are coveted by the continent and, in the case of European colonial states, operated as windows on the world (Aldrich and Connell 1992). Thus, in the case of the near island, those that are closest to a coastline and likely claimed, how can their terms be solely discussed while under the sociocultural and socioeconomic expectations of a dominant, mostly adjacent state? As such, the nearer the island to the mainland, the more subordinated their terms may become.

Before proceeding, a simple framework is offered to discuss these European islands or, furthermore, any island, as it represents how a piece of land surrounded by water undergoes gradations of complexity that are compounded by nearness. First, locate the island(s) and determine whether it is near or far, within an archipelago or alone. Second, discuss the type of island it is in the context of its nearness to a neighbouring landmass, considering first its physical composition, arrangement within its waterbody and orientation toward the coast. Third, with respect to climatic conditions, determine the physiography that has developed due to the geography of the island. Fourth, establish resource potentials resultant from the physiography and geography. Fifth, learn who uses and could use these resources. And, finally, sixth, discern who manages the resources, tangible and intangible, and their usage, and for whom.

The format of this chapter will follow the aforementioned framework. First, the concept of near islands and the divisions used to determine what is near and what is far are discussed. Europe is the foundation of this discussion, as an overwhelming majority of European islands are near. Next, we will discuss common island types with examples in Europe and how consistency between the mainland and islands particulars contribute to the perception that islands are an extension of the mainland. It is important to note, however, that most near islands are small islands and exist at the periphery by their very nature. This discussion will be followed by a brief and broad overview of the physical characteristics of European islands to engender an understanding of the scale of differences in European island spaces. In doing so, we delimit to countries with vast archipelagos or those with islands notably factored into their territory expansion narratives. This will include countries of the Fennoscandia Peninsula, Denmark, Scotland, France, Spain, Portugal, Italy and Greece, before ending in Croatia. While the entire United Kingdom would be a fantastic study of island-to-island relations and administration, many of the island territories in the United Kingdom are far or autonomous and thus outside of the scope of this study. Next, we will briefly discuss near island resources and their usage, two areas that explain mainland motivation and priorities. The following section discusses the demographic realities of European islands and how they inform the direction of the island. Finally, we will discuss different management schemes that have been used to govern the islands. This final point is crucial in this discussion of near islands, as it brings some clarity as to why the term *near* is far more complex than an explanation of distance but should also include the processes by which the mainland may see and seek to bring continuity between the mainland and island.

Despite island differences, some patterns will emerge. European states with near islands can be loosely arranged into three zones, north, continental, and south. Development in Europe tends to be higher in the north, as the countries locked within the Fennoscandia Peninsula had many periods of willing and compulsory collaborative growth depending on their status within the ruling regimes throughout time. Save the United Kingdom and, to a slightly lesser extent, Denmark, the northern countries were not as colonialist inclined, meaning their administrative and governing capacities were not tested across numerous outside spaces. As such, these countries were able to "hone" their respective frameworks across relatively homogenous land areas and populations, and, with that, learned to administer government across various land types.

Scotland, it should be noted, does not easily fall into this depiction with its islands being decidedly near enough to ensure consistency with the mainland but comparatively far-flung enough to have been possessed by Norway at different periods, which produced a blend of Gaelic and Norwegian cultures at sea. Additionally, Denmark was a colonial power as well despite its Scandinavian designation and commonalities of the north, while also holding many near islands; however, they will fall into the north zone.

The concept of intermingling cultures through territory expansion moves us directly into Continental Europe, where these states, along with the United Kingdom, are ground zero for colonial expansion and with that a more diverse set of land areas of which to adapt. The states of this set, France, Spain, and Portugal, and their total incorporated land and sea areas, did not materialize in an instant. They were the product of a gradual accumulation of tribes and kingdoms across the mainland throughout the centuries that eventually became individual republics. These states had to administer government across various land area types in various eras of territory change and their accompanying challenges. Most of these countries found their accumulated island territories in outer seas, the far islands that primarily make up the European Union (EU) Outermost Regions class, while their near islands are found along their southern, seaside borders. The remaining south states, Italy, Greece, and Croatia, find their near island story and accumulation of land area in the Mediterranean Sea, which is hardly a homogeneous region, although kindred due to climate, terrain and resources composition. Whether the geographies of incorporated land area for individual states directly impacts the near island and mainland relationship is not in focus here. Nonetheless, it does provide some points of delineation for the overall analysis.

To chronicle every near island or series of near archipelagos of Europe would likely be fruitless in furthering subject discourse. Such a narrative would also be an undertaking that would reduce *Les Hommes de bonne volonté* to a haiku. Instead, we envision this as continuing an introduction of near islands, set within the context of Europe, with an explicit focus on their realities and points of distinction among far and near islands.

PROFILING A NEAR ISLAND IN EUROPE

The term *near* is relative, although in this chapter I use it purposefully and with a clear intent. Geographic distance is, at first, simple to understand, although varying relevance to the term *near* withstands. Sicily is less than 5 km from the Italian coast, which is near enough to be bridged. The island of Hvar is 7 km from the Croatian coastline; however, we exclude the island of Sicily from this discussion. The Shetland Islands and the Balearic Islands are slightly more than 200 km from the Scottish and Spanish coasts, respectively. But this near island framing includes only the Shetland Islands. Area is also incorporated into this term *near*; however, exceptions are again prevalent. The island of Corsica is 8,600 km^2 and approximately 180 km from the French mainland. The island of Crete is 8,300 km^2 and approximately 120 km from the Grecian Coast. This definition of *near* will exclude Corsica and include Crete. Population and demographics must be considered in any island study; sparse populations, low population density, depopulation and seasonal inhabitants are prevailing trends on most European islands. The Isle of Mann is more than 500 km^2, approximately 70 km from the coast of the United Kingdom and has a population of more than 80,000 inhabitants; the island of Gotland is twice the size and distance from its parent state, with a population of roughly 60,000 inhabitants. Only Gotland is discussed in the context of the near island.

These examples may appear trivial, and that is intentional, as we seek to complicate the term *near* by adding another metric, autonomy. In the aforementioned cases, each of the islands excluded from our discussion of near islands has a designation of heightened self-governance granted to them by their parent state. The islands included in this definition of near do not. Mainland governing schemes capture them, and they are expected to conduct matters of administration and life within roughly the same framework as any other land area would, despite island specificities. Accordingly, near is also meant to define an island that can be reached by mainland mechanisms that embody the authority or general happenings of its parent state.

Autonomy is an undervalued component to island functioning. Island spaces are the definition of finite, and the components that shape their environments had to work in precise harmony to build and maintain the renowned biodiversity and species variation, and rarity, for which they are commonly known. Reshaping the land to suit the needs of an individual or unit was not the aim of the first islanders, nor did it guide anthropogenic activity from there on out. Instead, a series of phases can explain the islander–island relationship. The first near island settlers were likely not sailors, but farmers or some other land-based occupation, and first came to the island for survival, without knowing its attributes. In the next phase, these settlers had to develop an understanding of the land on which they arrived and accept the existing resources in their existing locations. Next, it can be assumed that islanders tried to utilise mainland cultivation skills and realised that the island does not "take direction," but instead sets the direction. Furthermore, we can deduce those islanders, through experiencing life on a limited land area with finite land resources and without full knowledge of the sea, moved toward living sustainably. In this scenario, the island–islander relationship can be explained as a gradual congruous outcome between the individual and the environment; therefore, islands are as sustainable as a minimum of resources would allow, and the inhabitants of islands had to respect and adapt to the limits of the island to ensure mutual survival. In this way, the introduction of the state or outside interests can act as a hindrance to island functioning. Thus, theoretically, only an islander, imbued with generations of conscious and unconscious habits that evolved into norms, could ensure mutual survival.

This dance between humans and the environment occurred in a vacuum of sorts. Disembarking onto islands happened well before the games of empire and state-building began, let alone before the forces of the Industrial Era and globalisation were unleashed. Hence, islander activity and customs could be extraordinarily space and place possessed, building the insular processes that contribute to the trait we refer to as islandness, another crucial component of island functioning. These dynamics occurred because the land and its inhabitants were isolated enough to have the independence to do so. While autonomy this full is not granted to an island that is not an independent state, what is granted with special designation is an acknowledgment of or at least deliberation upon the islanders knowing best. In the case of near islands, the geographic realities that could obstruct economic and environmental benefits to the state are acknowledged, but the islandness is denied, and with that the autonomy to preserve it, as in this framework, the modern state has decided that they know best because they are near enough to observe and administer.

Thus, we use this term *near island* to describe island spaces within the recognised seascapes or immediate maritime boundaries of the European continent, in the custody of a mainland state, and incorporated into the overall constitutional framework specific to state territorial management, absent any broad self-governing concessions. Under this depiction, these islands can be bridged or unbridged, small or large, sea-based or inland, or densely populated or depopulated.

ISLAND TYPES IN EUROPE

Discussing the physical composition and arrangement of a near island may appear trivial at first; however, these metrics set the foundation for state expectations. Continental and barrier islands are the most frequent in Europe. There are few volcanic islands in the central Mediterranean, most notably the Aeolian archipelago; several Grecian islands in the South Aegean Sea, including Santorini; and a small outer island in the Adriatic Sea, Brusnik. Tidal islands are comparatively, predominately found in the northern stretches of Europe, for instance, Lihou, one of the Channel Islands, or Le-Mont-Saint Michel in France. Europe also has a series of lake and fluvial islands, also known as river islands, in the northern states of Finland and Sweden, which account for the majority of Finnish and Swedish islands despite the countless islands that form their fragmented shorelines.

Despite this, virtually all near islands of Europe are continental islands, based on proximity, placement upon the same continental shelf as some mainland, and possessing similar terrestrial characteristics and physical composition to the nearby coastline. Near islands either broke away from the mainland during the formation of the continent thousands of years ago or were once a system of hills, mountains and valleys before deglaciation and subsequent sea level rise filled those spaces and created new seascapes. The latter point is the likely origin for many of the islands of the north and those of the more eastern parts of the Mediterranean. Barrier islands are more infrequent than continental islands and found along 6 to 13% of coastlines worldwide (Ulm et al. 2016). They are also linked to the same continental shelf, although they are sediment based. There are a few examples in Europe. The Frisian Islands, which stretch from the Netherlands to Denmark, are barrier islands formed throughout the millennia via sediment transport in the North Sea, which was once a dry land area before deglaciation 18,000 years ago. In the south, Lido di Venezia, off the coast of Venice, is a seven-mile stretch of land flanked by the Venice Lagoon and the Adriatic Sea.

Barrier islands are very close, almost parallel, to the coastline, and act as a protective buffer for storm waves and prevailing winds; however, so too do the near islands of Scandinavia, Finland, and Croatia. While these archipelagos have some islands with sedimentary characteristics, they are consistently a belted constellation of similar geology along the coasts of their respective mainland. Accordingly, these archipelagos have characteristics that could classify them as continental-barrier islands. This discussion is relevant because different island types determine perception by way of proximity, especially whether the island is treated as an extension of the mainland. It is reasonable to assume such because of island nearness in both distance and physical composition. In the most extreme cases, these islands come into mainland focus and are bridged, becoming an exact extension, begging the question already touched upon in the previous chapter: If the island is bridged, is it still an island?

There are several examples throughout Europe to consider. The larger near islands of Denmark have the same low-lying terrain and physiography as the mainland and provide passage to Sweden. What were once sea passages utilised by Scandinavian regimes are now bridges linking the Fennoscandia Peninsula to Continental Europe. Nearly all of the islands of the south and southwest Finnish border are bridged. Venice is known as the "floating city" when it is, in fact, a cluster of islands linked by small bridges. The island of Krk, Croatia, is one of the larger islands of its archipelago and approximately 5 km from its coast. There is a common perception throughout Croatia that Krk is bridged and not the same as other islands. Island types are somewhat clear in their particulars, and most likely geographic locations to draw some inference into what natural and anthropogenic processes can occur. The addition of the bridge alters this perception and aligns the processes on the island with mainland framing and expectations; therefore, bridged islands are simultaneously two things and thus a different type of island. While they are not considered a class within island types, the discussion for the inclusion of bridged islands is undoubtedly worthy.

This strikes upon a question of whether further classification for island types is necessary; however, this discussion of bridged islands says nothing of frequency, space and place. For example, islands in the Scandinavian states, Finland, and Croatia are all archipelagos that form a perimeter barrier between the mainland and the sea. The Scottish islands, the Shetlands, the Orkneys, and the islands of Eilean Siar are far removed from their parent mainland but are an arrangement of many islands of different size, i.e., an archipelago. But are they an archipelago in the same sense, or are they a clustering of islands? This is a distinction that requires more attention, as they tend to directly or indirectly influence how the state views and uses the islands.

WHERE ARE THE NEAR ISLANDS OF EUROPE?

A tour of Europe begins best with islands in the northern outer reaches, Iceland, an autonomous state, and the Svalbard archipelago, self-governing Norwegian territory. Further southeast, we reach into the semifrozen igneous and metamorphic rock landscapes of the Fennoscandia Peninsula, which houses Norway, Sweden and Finland, and thus the majority of European islands. Many of the archipelagos outlining the northwestern stretch of Fennoscandia, and much of the northern territory in general, belong to Norway and are located in tundra climates. Warm waters from the North Atlantic Drift and Gulf Stream, along with southern winds from prevailing westerlies, influence the entire climate of continental Europe and thus alter the climate from the south of the peninsula moving north, transitioning from subarctic to a climate more consistent with mainland Europe moving south. Deglaciation from several millennia ago and a buffering of southern wind and water activity to Siberian High winds give way to a fully habitable terrain comprised of a vast system of lakes and rivers weaving in and around a mostly flat and heavily forested terrain of contiguous land and interior lake islands. These features characterize the majority of Finnish territory, although also the inner south of Sweden. In both countries, interior lake islands outnumber sea islands (Källgård 2005). On the Norwegian side, earth-folding processes gave way to its mountainous terrain. Successive periods of glacier melt carved through dense hard-rock terrain, transporting rock, sand, gravel and other sedimentary materials to form the system of fjords and bays, as well as innumerable islands and skerries, that characterize its coastline.

The Fennoscandia Peninsula eventually splits in the south into the three distinct peninsulas before further fragmenting into thousands of islands in the northern Baltic, North and Norwegian seas. In the North Sea, the Shetland and Orkney archipelagos are off the extreme north of mainland Scotland. Each archipelago has the arrangement of a larger island with smaller satellite islands. The islands are contrasts in topography. The Shetland Islands are farther north and are a complex mix of metamorphic rock, Old Red Sandstone soils in the east and igneous granite rock on the west, and fertile limestone valleys, sweeping along a gentle hilly terrain. The island landscape has been altered through centuries of deforestation by way of grazing and burning. The Orkney Islands have a flatter terrain with jagged rock coastlines, similar to the northeast coast of Scotland. Deforestation has also altered the terrain throughout the centuries, with the aid of westerly winds battering the coastline and washing over the low-lying terrain. Old Red Sandstone is common throughout the archipelago, although there are also small deposits of visible limestone and granite. The combination of glacial deposits and sands have given way to the Shetlands' lush and fertile landscape.

A passage from the coastal city of Malmö, Sweden, tunnels beneath Øresund Strait and connects to an artificial island that bridges Sweden to Denmark. Denmark rests within a post-Variscan sedimentary basin that stretches from the Baltic states to the coastline of northern France (Neubauer 2003). Several periods of glaciation and deglaciation have culminated in producing a flat landscape of sands, clays and gravels that is not exceptionally fertile. Danish islands account for a substantial amount of its overall territory.

Rocky coast outlines the perimeter of the north Atlantic coast of continental Europe, including the states France, Spain and Portugal. These coastlines change in composition and elevation throughout, demarcated by a nearly even mix of hardstone outcrops, for example, granite and limestone, as well as such soft stone as chalk and colluvial–acolluvial sediments, with gradual entrances in the sea and high vertical cliffs (Gómez-Pujol et al. 2014). This portion of the continent is susceptible to hydrological forces of the Atlantic swell and northern and westerly winds, which make islands in the north far and few. There are some islands present, although with peculiarities. For instance, despite the temperate climate of Atlantic Europe, the Isle-de-Berhat has a Mediterranean microclimate and accompanying vegetation over a terrain of pink granite. The small island of Ushant has lush and fertile grounds despite its western location and vulnerabilities to wind and sea. Other islands outlining the French coast have a similar profile and oceanic climates over a geomorphological profile of the coast.

This dramatically shifts and transforms upon crossing through the Strait of Gibraltar and into the fabled climate and terrains of the Mediterranean Basin, a seascape that touches the shores of Europe, Africa and Asia. The geographic extent gives way to a highly varied topography and biogeography that ebbs and flows across a system of hilly and mountainous terrains that encapsulate the region from central and eastern continental climates, sandy beaches and rocky shorelines, all of which encompasses rich and rare vegetation, flora and fauna on the mainland and diverse marine life at sea. The Iberian, Italian, and Balkan peninsulas and their accompanied islands demarcate the European portion of the Mediterranean. The entire region has a characteristically Mediterranean climate, save for the northern stretch of the Balkan and Italian peninsulas, and the southern stretch of the Iberian Peninsula. Nevertheless, hot summers and cool, humid winters are synonymous with the region, which has had a profound effect on vegetation, wildlife and marine life.

The basin is a world biodiversity hotspot where endemism runs high, as the Mediterranean was isolated from the effects of the last ice age. Both terrestrial and marine life flourish here with little natural interference. While the Mediterranean scrub is most common, many other biomes, for instance, forests, woodlands and shrublands, are also abundant. Olive, fig and fruit trees are prevalent throughout the islands and coastlines, as well as various grape species, which contributed to the Mediterranean reputation in wine production. Plant life in this region tends to be denser, waxy plants that hold water and can exist without constant irrigation.

Geomorphology is also quite varied in this region. Despite the sandy shores commonly used in tourism promotion, the Mediterranean is quite hilly. The Iberian Peninsula is mountainous along its southern border but also sediment based in the areas that reach into the French coastline. The Adriatic was once a dry, mountainous terrain, contributing to the varied elevations found throughout the Croatian archipelago. Seismic activity at the boundary of the Eurasian and African plates formed the volcanic islands found primarily around the Italian coast and in the Aegean Sea. The stone cliffs and rocky shores of the Mediterranean mainland and islands are predominately such softer stone as karst, sandstones and limestone, although there are denser limestones to be found in the eastern portion of the basin, specifically in the Ionian Islands off the western Greek coast.

Marine biodiversity is exceptionally rich in the Mediterranean, representing 4 to 18% of global marine biodiversity despite being a small fraction of the earth's oceans. Natural marine pathways and processes have contributed to species richness throughout the centuries, as the basin rests along several intersection points: the Atlantic Ocean by way of the Strait of Gibraltar to the west, the Black Sea to the east and the Poe River to the north. The introduction of the Suez Canal to the south, accompanied with stresses from additional human activity, predominately shipping, and global climate change has gradually altered species composition, exacerbating the challenges stressing an already fragile seascape.

NEAR ISLAND INDUSTRY

The physical composition forms the foundation for island vegetation and biodiversity and development trajectory. In other words, the resources of the near islands afforded to the first islanders, the land that they could cultivate and arrange themselves around, the climatic circumstances in which they had to become accustomed and the range of life that had to subsist together. Additionally, they explain what put the island into the focus of the mainland or some other range of sea-explorer beyond its geographic location. For instance, the Fennoscandia Peninsula, while vast, is not topographically or climatically suited to be resource rich at all. Thus, the consolidation of resources was crucial to past occupying regimes and effectuated the intricate network of bridges connecting the islands and mainland, and thus the countries to one another. The flat, watery landscape of Finland does not allow for mass urbanisation in its island-abundant interior, the reason for its traditionally smaller populations, which pushed denser development to the coast; therefore, the lumber and milling industry became dominant and is still so in the Finnish interior. The denser southern portion of the country was more geographically convenient to reach, a necessity during its many times under occupation of the ruling, western regime. Furthermore, its islands are of utterly separate geology and climate than the interior, which may have allowed for greater options in physical planning and socioeconomic development.

In the Mediterranean, resource culling was equally important, although the practice tended toward the convenience of the noble and ruling classes. Islands of the south, although near in many contexts, are more widely dispersed in comparison to the north, making location a crucial factor. Ruling empires could rely on an island to be a resource extraction hub, as well as a key added geostrategic point within its maritime boundaries. Today, this holds but in the purpose of pursuing tourism. Europe has a long, captivating history, and islands are great stewards of old civilisations and customs that arranged themselves around their immediate environment before matters of the state became a part of the islanders' responsibilities.

The rich profile of rare and endemic terrestrial and water species are the most apparent attributes to draw in tourism, although their origins are rooted in island-sustainable functioning. Near islands do not face the pressures of obligatory sustainable functioning as far islands, as the latter group is too distant to be captured by mainland networks. Near islands, on the other hand, can be prone to distortion of sustainable living under mainland networks, even to the point of extreme specialisation in agriculture, military industry, or, in present trend, tourism.

"Backwards" is the description most given to the European near islands and other remote regions with systemic structural constraints that inhibit growth. Low GDP per capita, higher costs of living due to high insularity, small-sized markets, weak competition, low diversity in sectors and poor connectivity are just a few of the challenges islands face (Moncada et al. 2010). Islands also tend to have remarkably weak infrastructure when compared to the mainland and more natural barriers to development that require alternative solutions, although islands rarely rise enough in importance to become sites of innovation investments. Consequently, island economies tend to be a mix of traditional industries like agriculture and fishing, nonmetal mineral extraction, fishing and shipbuilding, and more modern service industries to accommodate tourism. High environmental values and contributions to state cultural heritage make islands attractive destinations, and islands throughout the continent are seeing an increase in visitors and, thus, human activity. Tourism is actually displacing traditional industries on islands throughout the world. In Europe, specifically the south, this is a problem, as the displacement of traditional industries means the weakening of sustainable functioning of the island.

The tourism industry has varying presence on all inhabited European near islands, but not all islands are tourism dependent. In the north, specifically, the islands of Finland and Norway benefit from their locations in and around natural resources hubs, natural gas and oil, and lumber, respectively. A robust services industry, from which its mostly bridged islands undoubtedly benefit, powers the Danish economy. The issue of tourism overtaking traditional industry appears to be a continental and south issue. Geography helps explain this. Many of the northern islands are bridged and within strong economies with ample social services and higher levels of local authority. The continental near islands are low in frequency and small in size, typically attached to the mainland administratively and thus likely dependent on the mainland economically. Although the southern islands are more numerous than continental, the size and frequency of these islands, poor connectivity and general characteristics of backwardness culminate to reveal low-competitive, low-innovation spaces (Moncada et al. 2010). Agriculture is not a robust sector on many near islands, as most plots are small and fragmented, and require fine coordination among islanders to yield a profitable product. These spaces are also too small to host mineral resources vast enough to extract on a full scale. Connectivity among near islands in these two zones is poor because links via trade are weak or nonexistent, as the spaces and their resources are mostly homogenous, at least in larger archipelagos with denser arrangements. The link, and with that the desire for traditional industries to be profitable, are found with the mainland.

A position within two seas and an ocean, and two climate zones with wide-ranging biodiversity values, power the fishing and marine-based industries of Norway. The Western Isles of Scotland are equally strong in fishing and shellfish and fish farming, having a dominant position in comparison to the other Scottish islands, offsetting the poor resource diversity on the islands. Most islanders looked to the sea eventually and understandably; however, these industries are threatened as well. Climate change and transport and trade routes present the most pressing challenges to near islands because they provide pathways for species composition alteration. Balance within an island space and islanders protective of the environment on which they depend determines the quality and abundance of the marine environment. Additionally, near islands are close enough that, contemporarily at least, fishers have to compete among one another in marine territories, which can produce market distortion of marine-based products. This is especially so in states like Croatia, where the islands are densely arranged. The competition can result in overfishing, which is an indisputable reality in the Adriatic from both the Italian and Croatian states. Moreover, when the market does not favour the trade and the seas are changing what is traded, the result can be the abandonment of the industry, and, next, the island.

Accordingly, boosting and regulating agriculture and fisheries activity is of high priority for the EU, with various funds designed for this reason. The regulation of the sea is of incredible importance to near islands for the aforementioned reasons and to stem the adverse effects of mass tourism.

Land-based tourism threatens the sea as much as the land. The anticipation of tourism and its immediate gains lead to overdevelopment of the coasts or land composition, which will impact the surrounding sea space. Sea-based tourism, for instance, sun and sea tourism, yachting, boating, beach creation and the like are directly affecting sea spaces; however, the design of policies to boost traditional industries on islands, in their inception, did not adequately consider near island handicaps. While EU island policy has been refined, challenges persist, namely, liberalisation of transport networks that were not curtailed to address the inherent connectivity issue of near islands, especially within archipelagos with high island frequency. As such, the changes and creation of EU policy toward traditional industries have been incremental, reacting to revelations of difficulties, as opposed to a comprehensive design that is more reflexive.

NEAR ISLAND DEMOGRAPHICS

Island inhabitants "can be" ascertained at the EU and state levels. There is an emphasis on *can be* because it depends on the population range used to determine administrative subdivisions of the state and how that moves upward to EU statistical counting. The EU NUTS framework is not built to account for near islands (Moncada et al. 2010). NUTS classifications 1 through 3 demarcate state, regional and territorial divisions of local administration specific to the state, usually a county or department, using population range for delimitation. Near islands rarely rise to the NUTS level unless they have large populations and are usually captured at the LAU classification, which EU statistical counting does not include. There are exceptions, for example, Gotland and Zealand in the north, and Crete in the south. In states where island territory significantly contributes to overall land area, more near islands are captured within a NUTS class, the best examples being Denmark and Greece; however, this is usually the satellite island within a small island cluster. Accordingly, near islands typically fall within a unit that includes other islands or the mainland. EU NUTS areas and state administrative divisions usually harmonize, so the former scenario is helpful in that islands are accounted for, along with other islands at the NUTS 3 level. This indicates that the islands are in direct focus and being managed by another island with original insight on island-specific issues and characteristics. The latter arrangement is a more problematic scenario. To include an island in the mainland creates a dynamic where the island population is likely as attended to as they contribute to the mainland unit to which they are administratively attached. The examples in Table 2.1 show the islands that rise to the NUTS 2–3 category. What is not apparent, however, is that in each of these cases, there are several to numerous near islands. Thus, the classification does not fit over an individual space as it does on the mainland; instead the classification blankets over a fragmented administrative set.

Table 2.1. Islands Reaching NUTS 2 and NUTS 3

Island/Archipelago	*Country*	*NUTS Level*
Gotland	Sweden	3
Eilean Siar	Scotland	3
Orkney Islands	Scotland	3
Shetland Islands	Scotland	3
Zealand*	Denmark	2
Bornholm	Denmark	3 (incorporated in NUTS 2 Zealand)
Fyn	Denmark	3
North Aegean Islands	Greece	2
South Aegean Islands	Greece	2
Ionian Islands	Greece	2
Crete	Greece	2

Source: Eurostat (2019).
Note: * indicates island with nation capital.

Regardless of the arrangement, theoretically, the census process of a state is designed to account for its territory, so their framework is built to reach into the municipal level and estimate the population. But there is another complication. Near island populations are difficult to estimate because the officially registered population on the island may not be the actual, permanent population (Källgård 2005). While near islands have their challenges, there are such convenient advantages as lower transportation costs for mainland-to-island travel and rural area subsidies and tax relief, as well as subsidised ferry parking, that impel residents to be administratively accounted for at sea while living on the mainland. How official versus actual residency is regulated depends on the state and if it is regulated appears to depend on how disruptive it is to state or local government revenue collection and disbursement. Additionally, to calculate the amount on the EU scale has proven to be very expensive, meaning the cost for weaker economies would be additionally burdensome (Armstrong et al. 2012). What is consistent with near island communities regardless of the actual population is that they tend to skew older, have less technical and computer skills, and have lower education level attainment.

Near islands are also depopulating. Presently, there is no higher risk to islands than negative population trends. Island populations tend to be of advanced old age, and younger people are leaving such island-centric industries as agriculture and fishing, and heading to the mainland in search of higher incomes and diverse social conditions. This trend is not entirely consistent throughout Europe. Northern islands have the benefit of the shared history of the Scandinavian states and Finland, and these near islands are better connected via bridges and tunnels, giving island natives better access to opportunities if their island does not present them. Norwegian islands are bolstered by long-running strength in fishing and sea activities, as well as petroleum and natural gas exploration, where the island location is a comparative advantage. Additionally, these are states with somewhat liberal migration policies used to stem depopulation trends on the mainland. How much migration policy affects the islands is unclear, as it depends on the location of out-migration and the local level capacity to accommodate migrants without island-specific skills.

In Greece, there is a security need to keep these islands populated because of shared history in the Aegean with neighbouring Turkey, and the country has moved policy to ensure they incentivise staying on or relocating to the islands. The tourism economy has presented a compelling reason to remain on the Greek islands, although the high seasonality of the sector presents a new set of demographic and census accuracy challenges (Moncada et al. 2010; Spilanis and Vayanni 2004). Depopulation and the economy of the islands are linked. As mentioned, near islands have backward characteristics that impede the development of diverse industries. Without diverse industries, there is low diversity across many demographics, specifically age, sex and education attainment. These trends are highly characteristic of southern near islands, as south European states are experiencing negative demographic trends, with the younger and more educated population leaving for city centres or heading north where economic opportunities are far stronger. Croatia is a prime example with island populations of advanced age, a growing diaspora and worries of a substantial brain drain (Nejašmić and Mišetić 2006). As younger, educated people leave the city centres, opportunities open for islanders, draining the island of younger generations to continue in traditional industries; this effectively causes harm to sustainable island practices, as there are fewer people with which the knowledge can be shared.

Mitigating this pattern is rife with issues, many already mentioned. Data collection of island demographics is the most obvious. In the cases of high island frequency archipelagos or clustered archipelagos, like those found in Greece and Scotland or Italy, respectively, enumerating the population and its characteristics is somewhat easier, as the islands are large enough to be at the municipal level at sea or administratively tied to another island. In the case of continental islands, specifically those of France, the near islands are too small and underpopulated to be administratively demarcated at the municipal level, and their populations are accounted for along with the nearby mainland. This is also present in the administrative arrangement of Italian near islands that do not benefit from being of a clear archipelago or cluster, for instance, the Tuscan or Aeolian archipelagos, and are tied to not only the mainland, but also the nearest main city centre of the mainland. Croatia has the dubious honour of housing every present arrangement within one archipelago, where the islands can be a single administrative unit, tied to another island, or tied to the mainland. The key takeaway is that while it is clear that islands have adverse demographic trends, the totality of extent or impact can be difficult or possibly impossible to ascertain.

NEAR ISLAND MANAGEMENT

We have touched on the physical and socioeconomic characteristics of islands; however, the question is, how is all of this being managed? In the cases of near islands, what is certain is that each island conforms to the management scheme of the state—meaning that if the island is a municipality or incorporated area, the same processes of any other land area in the state also govern them. From this point, each state uses a series of mechanisms to further nuance the position of the state concerning the island. Most near islands are too small to be captured at any administrative level higher than the local unit, which can include communes, parishes, municipalities and the like. So, the influence that a local population, islanders, have on the management of their land area depends on how much strength the state bestows upon the local unit. For example, in Finland and Scandinavia local governments are exceptionally empowered. Finland is a series of self-governing municipalities that has broad competencies in taxation, security and planning. In contrast, Croatian municipalities are not given broad powers to chart their own direction and are heavily dependent on the county and the state, making reflexive or reactive measures in development nearly impossible. In fact, one island may contain various municipalities with different party leadership or be administratively tied to the mainland, furthering the complexities.

There is no consistent island governance approach throughout Europe. Nevertheless, some patterns develop based on geography and frequency. First, the islands of Finland and Scandinavia seem to be captured within the overall governing policy of the state, meaning that islands are treated like any other area. No specific island policy is found here, save for the Island Act of Finland, a document that seems to formalize that islands are distinct areas, although there are no specific island prescriptions. Regardless, the islands of the far north seem to benefit from this arrangement, as the north is more decentralized than Central and Southern Europe, meaning that local governmental units have broad powers and thus greater autonomy to respond to island-specific issues. As such, local island units can chart their course within the state with more latitude. Local units tend to be clustered together in the north and not administratively attached to some mainland city. As such, islands tend to be administered by islands, which assumes more nuanced directives and governance given the insight.

Moving west, and into Scotland and the rest of Europe, discussions of island autonomy guide the administrative framework. Each country seems to be using policy to either demarcate their near islands as special territories or use policy mechanisms and governmental institutions that acknowledge and respond to island specificities. How much autonomy is necessary appears to be the overall question. It can be rightfully assumed that the greater the distance, the more socioeconomic and cultural distance the island will have with the mainland. In areas where there is a mix of histories with neighbouring countries, there are more reasonable claims to the territory. For example, consider Greece and Turkey and Scotland and Scandinavia. In both of these cases, islands are numerous and account for a significant part of the overall territory. Because these islands are the scene of some back and forth between nations that possess or once possessed them, policy mechanisms are aimed at granting autonomous concessions to keep them incorporated into the state. While not considered a near island by any metric, the Faroe Islands tend to be a guide for most of these areas, as their status within Denmark is tenuous without the sweeping autonomy afforded to the archipelago.

The Scottish islands have made moves toward decentralisation in the vein of the Faroe Islands but were granted a compromise instead. This is an area that has to continually be managed by the Scottish state with two factors in mind, literal and social distance. The metric of real distance is understandable; a state is far away and sometimes unable to craft policy that targets the islands efficiently. But it is the social distance that is more difficult to manage. Literal distance can be remedied in some ways by use of parliamentary representation and enhanced connections; however, if the islanders feel entirely culturally detached from the state, then their appetite for full detachment is stronger. The north islands, Orkney and Shetland, have been contested areas between Scotland and Norway, which has produced an evident cultural mix of Gaelic and Norwegian customs that distinguish them from the mainland and thus separates them from the mainland. Accordingly, there have been numerous attempts for autonomous designation within the Scottish state because the nuances to their living and customs are best managed directly. Scottish island policy has changed and adapted during the last

50 years to respond to such. One of the most recent iterations has been the creation of an "island-proofing" mechanism that qualifies legislation against a set of island indicators.

Greece is similar to Scotland in this regard. Grecian islands are located in the Aegean and Ionian seas before Crete spills into the greater Mediterranean, where the sea and surrounding lands are arguably the most contested in history. The borders of the mainland states were remarkably fluid into the 20th century and with that the island territories off their coasts. Islands account for 17% of Greek land area and more than 70% of the maritime borders. Greek and Turkish maritime borders are directly beside one another in the vein of a contiguous land area, with the islands determining their placement. Accordingly, Greece has a rather complex system managing their islands and keeping them populated, seemingly arranged around 1940 legislation, Law 2344/1940, which provides explicit definitions for its coastline and islands, and their intended use. Additionally, Greece was the only state with a dedicated ministry to its islands, making island oversight a cabinet-level priority. Since, however, that ministry has been collapsed, and the responsibility of island management has been dispersed throughout several ministries.

Greece and Scotland are examples of highly nuanced island policy and oversight. Other countries of interest here do not have as many islands or their place within the state cannot be reasonably contested; those that can are autonomous, for instance, the Faroes, the Aland islands, and the outermost regions of Spain and Portugal. Croatia is an odd outlier in this regard. Croatia has thousands of islands, but they do not contribute much to the overall land area of the country. Several different empires and states have claimed them but are nestled too close to the defined borders of Croatia, regardless of their time as incorporation as a territory to an empire or federation or as an independent state. Nonetheless, their island management does not reach the level of nuance found in Greece and Scotland. Whether the history of possession in a cul-de-sac water body is a factor is indeed worth investigation; however, Croatia does have an island act to manage its island development, and it is a sprawling document in comparison to that of Finland. The other European nations have much fewer islands than that of Scandinavia, Finland, Greece and Croatia. As such, they operate as Scandinavia, in that they are captured into the mainland territory management scheme. But they are rarely numerous enough to be administered from the sea and are usually attached to a mainland municipal-level unit. The exception is the Italian archipelagos and near islands, which are not as numerous but managed by sea and the mainland.

On the whole, European islands have weak local economies, and EU island policy is only recently slowly coming into form. Islands have to follow the governing framework of their parent mainland, so the EU policy communicates strategy direction toward 2020 goals more so than governing direction. How islands adapt to this strategic direction is discretionary. Overall, the foci of EU policy are wrapped around creating decarbonized economies and the *smart* island concept, which is an island that has achieved sustainable local development in demographics and economic terms that result in higher life quality; however, as mentioned, EU strategy and focus can only cover islands listed at the NUTS 2 or NUTS 3 levels, which does not extend down to the vast majority of near islands (Armstrong et al. 2012); however, the European Regional Development Fund (ERDF) and Cohesion Fund (CF) are designed to capture and disseminate to islands and other remote regions. A characteristic of EU funding and policy toward islands is that they fall under some remote region or geographically handicapped description, which includes mountainous regions and rural areas. Studies have also shown that the geographic handicaps of the latter two areas are not as constraining as islands. Without a specific focus on near islands, the use of these funds depends on the efficiency of the state in addressing island issues.

The EU has outlined supporting the islands in its treaty and various policies that have followed, citing the need to correct development imbalance areas that are geographically handicapped. Currently, without the ability to directly regulate the near islands within the bloc, incentivising with structural funds has primarily been the EU policy apparatus used to bring territorial cohesion with the mainland. The main structural funds for islands are the ERDF and CF, and Social Funds (ESF) and Agriculture and Fisheries Funds (AFF). Each brings a different element to help alleviate the effects of natural handicaps of island territories.

Research shows that the efficiency of the ERDF funds depends on how well a country is mitigating island-specific characteristics or, better, insularity. First, as previously stated, EU funds were designed to service larger territorial units. Near islands at the LAU level will thus enter into the funding scheme competing with their counterparts at land and sea, in the NUTS 3 category to which they are incorporated. Geographical handicaps and weak connectivity of an island will likely result in projects that are reactive in nature to mitigate insular factors from the onset, before any discussion of niche or innovation. The demographic profile of an island is one that does not support growth beyond niche markets, and these markets are not efficient throughout time. As such, these funds are as efficient, as the island is serviced by its governing administrative unit. In this way, how a country is able to service its island communities provides insight on how effective EU funds will be.

The CF and ESF directly address key island issues, which is reducing economic and social disparities; however, neither make any mention of islands in their language. The prerequisites of environmental protection and sustainable development in the CF suggests compatibility with islands. ESF are made to service more economically vulnerable communities and are dispersed based on the overall wealth of the state, making them attractive funding avenues for the southern near islands. AFF cofinances member state projects that help fishermen transition to sustainable fishing, support economic diversity in coastal communities and improve quality of life along European coasts. The newest iteration of the AFF, for the 2014–2020 programme years, mentions islands specifically, as is expected. Paragraph 82 of the Marine and Fisheries Fund (MFF) outlines that its funding and regulations should be complimentary to islands and island states at national and subnational levels, provided the funds are allocated toward sustainable development and enhanced maritime cooperation between member states. Additionally, the paragraph encourages the intersection with other EU structural funds; however, each of these funds have the constraint of the ERDF; they are unable to be accessed beyond the NUTS 3 level.

Before delving into deep scrutiny or criticism of EU funds, I must admit that this analysis is somewhat superficial, accounting for the process and structure of funds without acknowledging the various mechanisms each state will have to procure these funds. As has been continually stated, each near island has its own specificities. To determine the successes and shortcomings of the EU structural funds in addressing near island needs would require a deeper investigation into each archipelago, which is outside of the scope of this chapter. Nonetheless, what should be taken away is that EU funds were designed to include islands as remote areas, without broad confirmation or mitigating criteria of particulars that differentiate them from other remote areas.

CONCLUSION

This chapter is a discussion of the near island situation; however, in honesty, islands are deeply challenging to generalize, so it barely scratches the surface. It is impossible to neatly encapsulate the near island situation into a single definition or classification of any type. Some things are indeed consistent. Near islands are largely misunderstood and rife with challenges because of that. An even clearer picture of these challenges and their implications would emerge through a deeper analysis of the archipelagos that I have discussed. Demographics, low-innovation industry, poor connections and even what to classify an island as in both a definitive and administrative sense are part of the day-to-day near island experience. How this is best managed and by whom is the prevailing struggle.

Autonomy, or specifically, how much is allowed, is the primary negotiation between near islands and the mainland. Near islands, in comparison to their far-flung counterparts, do not have the benefit of distance to ensure that their environments and cultures prevail on their own terms and must find ways to elevate their individual voices to the state. This is a struggle for any near island that is found within a vast archipelago and administratively tied to a neighbouring coastal city, a classification that most European near islands find themselves in. Island types are an underdiscussed metric despite being the likely initial starting point for this negotiation. What this means is that a difference of physical composition between near islands and the mainland will push the mainland to adapt to that specificity, with tidal and volcanic islands being the most obvious examples because they have the most contrasting composition and situation; however, tidal and volcanic islands are scarcely found in Europe even when their frequency with barrier and continental islands are not compared. When literal distance is its most near and composition is most consistent, the island is under pressures to be bridged, entirely altering its classification as an island.

Europe was used as the case for this discussion of near islands with good reason. It is a peninsula continent and stretches from the Arctic to the Mediterranean. A journey along its seascapes unveils a terrain weaved of hills, valleys and plains where glaciers and metamorphic rock transition into sediments and then karstic stone, where tundra turns temperate. There are many islands to be found along this journey, some archipelagos inhabiting various climate or geomorphological zones. As such, no two islands are the same, although that idea can be lost.

Demographic challenges are common on near islands and not simply because many near islands are losing crucial cohort groups needed for continuing a population. It is also because of discrepancies between recorded population and actual population. Census methods vary from state to state in Europe, and often the smaller near islands are counted within a cluster, obscuring the population totals in counting method. EU statistical counting methods do not always account for near islands, as they typically fall within the LAU levels. The two latter points are most detrimental for states with numerous islands, for example, Croatia, the Scandinavian states and Finland, and Greece, and its mitigation can only occur at the state level, should the state feel the need to elevate island demographic counting in their census.

How islands are managed is a study in contrasts. Here the concepts of autonomy and oversight greatly vary. No near island is offered autonomy at the level of such archipelagos as the Faroe Islands or islands within the EU outermost regions. How well the island is able to chart its own course depends on how devolved the state government is and whether the islands are numerous or large enough to be administered at sea, as exemplified by Finland and Greece. On average, islands are administered by a neighbouring coastal city, and how well their needs are managed depends on how many additional levels are above the initial level at which the near island finds itself. The closer to the bottom, which they often are, the harder it is for the needs to rise to the top. States with islands that greatly contribute to territorial sea and overall territory tend to have more refined policies for near islands or that account for near island importance, for instance, Scotland and its island-proofing mechanism and Greece with its ministries and progression of laws governing its vast archipelago. The EU has been moving toward an island policy, but as of now there is no coherent overarching policy, and whether such a policy would reach to the near island level is unknown. As with statistical counting, it is an issue of levels, as near islands are often at a level that does not allow for them to bid for crucial EU structural funds while having the development issues that arguably demand them most.

The notion, even legend, of an island can be powerful, conjuring images of an idyllic, temperate paradise untouched by time or the hallmarks of tumultuous mainland life—the "away" that people rush to get. This is ironic, as this is an imported colonial construct that had little to do with the European experience beforehand. Nevertheless, beyond the paradise lore are small, isolated spaces and inhabitants surviving within the bounds of their environment despite the pressures of itself and the neighbouring mainland, each different, each unique, and, despite that, a broad governing concept still perpetuates throughout nearly every European archipelago.

We are now to contextualize many of the themes discussed here to the Croatian islands, an archipelago occasionally referred to as the "Mediterranean in small." We will first emphasize the geographic and physical properties of the Croatian archipelago, as they have allowed life to flourish for centuries and aided in catalysing the initial disembarkations of the first islanders and those that followed. The past and present situations in the archipelago will also receive general attention, with explanations of how settlers coexisted with the island itself before later managing the pressures and expectations of the islands being gradually incorporated into wider territorial networks.

REFERENCES

Aldrich, Robert, and John Connell. 1992. *France's Overseas Frontier = Départements et territoires d'outre-mer*. Cambridge, UK: Cambridge University Press.

Alestalo, Matti, and Stein Kuhnle. 1987. "The Scandinavian Route: Economic, Social and Political Developments in Denmark, Finland, Norway and Sweden." *International Journal of* Sociology 16, no. 3/4: 3–38.

Armstrong, Harvey W., Benito Giordano, Thanasis Kizos, Calum Macleod, Lise Smed Olsen, and Ioannis Spilanis. 2012. "The European Regional Development Fund and Island Regions: An Evaluation of the 2000–06 and 2007–13 Programs." *Island Studies Journal* 7, no. 2: 177–98.

Baldacchino, Godfrey. 2013. "Island Landscapes and European Culture: An 'Island Studies' Perspective." *Journal of Marine and Island Cultures* 2, no. 1: 13–19. doi:10.1016/j.imic.2013.04.001.

Baldacchino, Godfrey. 2007. "Islands as Novelty Sites." *Geographical Review* 97, no. 2: 165–74. doi:10.1111/j.1931-0846.2007.tb00396.x.

Baldacchino, Godfrey. 2004. "The Coming of Age of Island Studies." *Tijdschrift Voor Economische En Sociale Geografie* 95, no. 3: 272–83. https://doi.org/10.1111/j.1467-9663.2004.00307.x.

Bergman, Ingela, Anders Olofsson, Greger Hörnberg, Olle Zackrisson, and Erik Hellberg. 2004. "Deglaciation and Colonization: Pioneer Settlements in Northern Fennoscandia." *Journal of World Prehistory* 18, no. 2: 155–77. doi:10.1007/s10963-004-2880-z.

Beriatos, E., and M. Papageorgiou. 2011. "Maritime and Coastal Spatial Planning: The Case of Greece and the Mediterranean." *Sustainable Development and Planning V*. doi:10.2495/sdp110011.

Depraetere, Christian. 2008. "The Challenge of Nissology: A Global Outlook on the World Archipelago." *Island Studies Journal* 3, no. 1: 17–36.

Eurostat. 2019. NUTS—Nomenclature of Territorial Units for Statistics. https://ec.europa.eu/eurostat/web/nuts/nuts-maps.

Fernandes, Ruben, and Paulo Pinho. 2017. "The Distinctive Nature of Spatial Development on Small Islands." *Progress in Planning* 112: 1–18. doi:10.1016/j.progress.2015.08.001.

Gómez-Pujol, Lluís, Augusto Pérez-Alberti, Ramón Blanco-Chao, Stéphane Costa, Mario Neves, and Laura Del Río. 2014. "The Rock Coast of Continental Europe in the Atlantic." *Geological Society, London, Memoirs* 40, no. 1: 77–88. doi:10.1144/m40.6.

Hall, Michael C. 2012. "Islands, Islandness, Vulnerability, and Resilience." *Tourism Recreation Research* 37, no. 2: 177–81.

Källgård, Anders. 2005. "Fact Sheet: The Islands of Sweden." *Geografiska Annaler. Series B, Human Geography*, Special Issue: Islands: Objects of Representation 87, no. 4: 295–98. https://www.jstor.org/stable/3554389.

Magaš, Damir. 1996. "Croatian Islands: Main Geographical and Geopolitical Characteristics." *Geoadria* 1, no. 1: 5–16.

Moncada, Stefano, Marguerite Camilleri, Saviour Formosa, and Roberta Galea. 2010. "From Incremental to Comprehensive: Towards Island-Friendly European Union Policymaking." *Island Studies Journal* 5, no. 1: 61–88.

Mueller-Dombois, Dieter. 1992. "The Formation of Island Ecosystems." *GeoJournal* 28, no. 2. doi:10.1007/bf00177244.

Nejašmić, Ivo, and Roko Mišetić. 2006. "Depopulation of Vis Island, Croatia." *Geoadria* 11, no. 2: 283–309.

Neubauer, F. 2003. "Geology of Europe". In *Volume Geology. Encyclopedia of Life Supporting Systems*, edited by B. De Vivo, Grasemann, B., and Stüwe, K.: Oxford, UK: UNESCO Publishing-Eolss Publishers.

Phillips, Tim. 2004. "Seascapes and Landscapes in Orkney and Northern Scotland." *World Archaeology* 35, no. 3: 371–84. doi:10.1080/0043824042000185775.

Ratter, Beate M. W. 2018. "Cultural History of Islands." In *Geography of Small Islands: Outposts of Globalisation*, 61–91. Cham: Springer International.

Royle, Stephen A. 2001. *A Geography of Islands: Small Island Insularity*. Routledge Studies in Human Geography. London and New York: Routledge.

Spilanis, Ioannis, and Helen Vayanni. 2004. "Sustainable Tourism: Utopia or Necessity? The Role of New Forms of Tourism in the Aegean Islands." *Coastal Mass Tourism*, 269–91. doi:10.21832/9781873150702-015.

Starc, Nenad. 2002. "Managing Island Development: The Croatian Case." *Sociologija Sela* 39, no. 151/154: 15–36.

Taglioni, Francois. 2011. "Insularity, Political Status and Small Insular Spaces." *Shima: The International Journal of Research into Island Cultures* 5, no. 2: 45–67.

Ulm, Marius, Arne Arns, Thomas Wahl, Steven D. Meyers, Mark E. Luther, and Jürgen Jensen. 2016. "The Impact of a Barrier Island Loss on Extreme Events in the Tampa Bay." *Frontiers in Marine Science* 3. doi:10.3389/fmars.2016.00056.

Wygal, Brian T., and Stephan M. Heidenreich. 2014. "Deglaciation and Human Colonization of Northern Europe." *Journal of World Prehistory* 27, no. 2: 111–44. doi:10.1007/s10963-014-9075-z.

Chapter Three

The Croatian Islands

An Introduction

Anica Čuka and Josip Faričić

THE ARCHIPELAGO

The Croatian archipelago contributes to the small islands of Europe with 1,246 islands, islets, rocks and reefs. It is a place of multiple contacts between the sea, karst relief, Mediterranean climate and diverse flora and fauna, where there are few freshwater sources or surface watercourses. In spite of its meagre resources, the archipelago has been inhabited more or less continuously since prehistoric times.

Situated in front of the Croatian coastline with which they form a geological-geomorphological unit, the Croatian islands were created geologically in the same way as the neighbouring mainland and were an integral part of it until the last Würm glacial age. Afterward, due to global warming and the resulting rise in the sea level, they were separated from the mainland. In the global and European context, the Croatian islands are really near islands. This label is due to not only their geographic proximity to the mainland, but also their socioeconomic fusion with the mainland regions and their coastal centres.

Nearness is a complex spatial variable and can be viewed diachronically and synchronically. The diachronic view assumes the historical developmental perspective according to which the nearness (or perceived nearness) of the island to the mainland and other islands was affected most by the degree of socioeconomic development, efficiency of administrative mechanisms, traffic connections and infrastructure. Along with socioeconomic development, the establishment of a traffic system, including increasingly fast and frequent links, the introduction of a modern democracy in local and regional self-government and the establishment of infrastructure of increasing quality and coverage linking the islands with the mainland led to decreasing dependence on natural factors (particularly weather factors) and increasing functional convergence between the islands and the mainland. The general rule is vividly illustrated by negative examples. The Croatian islands, which have no regular passenger lines and are not integrated in the electricity supply network of the neighbouring mainland, have remained completely isolated and lost their population. This happened several decades ago to the islands of Škarda (3.3 km^2) and Sveti Andrija (4.2 km^2), where for centuries there had been small island communities that had built a self-sufficient economy to meet their modest needs.

The synchronic view refers to a certain point in history when various factors correlated to contribute to the nearness of the islands. The main factor (although not always the deciding one) is geographical distance. This is particularly highlighted if an island is joined to the mainland by a bridge (Krk, Pag, Vir, Murter, Čiovo and Trogir) or, to a lesser extent, if the island is close to a mainland centre. For example, Preko, on the island of Ugljan, is 2.7 nautical miles from the mainland city of Zadar, a distance that can be travelled by boat or ferry in 25 minutes. During the summer season, there are 29 crossings per day, and 25 out of season, which is three times more than the public transport buses that link the historical centre of Zadar with the outlying suburbs. On the other hand, the island of Dugi Otok is 18 km as the crow flies from Zadar. The nautical route from Zadar to Brbinj, halfway between the SE and NW ends of Dugi Otok, is 24 km (13 nautical miles). The island of Rivanj is 15 km from Zadar as the crow flies, but the nautical route is 19 km (10.3 nautical miles). It is actually closer to Zadar than Dugi Otok but is smaller and served by only one fast passenger boat a day (35 minutes) and two ferries a week (1 hour and 45 minutes). Dugi Otok is served by a fast passenger line (45 minutes) four times a day and a ferry (1 hour and 40 minutes) three times a day. The frequency of its connections with the mainland regional centre makes Dugi Otok closer to the mainland and shows how island accessibility is an important variable in socioeconomic interactions. If a fire or shipwreck occurs, or if urgent medical intervention is needed in response to a heart attack or stroke, for example, the accessibility of an island grows into a variable that can mean the difference between life and death.

Many Croatian islands are quite distant from the mainland in the socioeconomic sense, no matter how geographically close they may be. Today, this primarily means the islands that have no regular passenger lines or at least suitable harbour infrastructure, those that are not covered by telecommunications network signals and those that are not linked to the mainland water supply (a large majority of Croatian islands do not have their own natural sources of drinking water). Most inhabited islands are linked to the mainland electricity supply network, but their connections are old and wearing out, so that there are often power cuts in bad weather, paralysing most island activities.

The inhabited Croatian islands, apart from the bridged ones, are cut off from the mainland for several days each year due to bad weather. High winds and seas interrupt nautical passenger routes to and from the mainland, and insularity becomes pronounced. Since the islands rely on towns on the mainland to a great extent, the islanders find such interruptions difficult and think their lot is worse than that of the mainland inhabitants, especially those in the towns and cities. At such times, all the islands, regardless of their proximity to the mainland, become remote islands.

TOPIC, METHODOLOGY AND RESEARCH AIM

The main topic of this chapter is the Croatian islands as a geographic unit. By analysing and correlating various sources of spatial data, from archive records and maps to the latest statistics and field observations, and with reference to the existing literature, the basic spatial processes and relations on the islands situated in front of the Croatian coastal region can be affirmed. To this end, a geographic matrix was researched, which has focused in different ways on various social and economic processes and spatial interaction, and which has only changed inasmuch as the islanders themselves have changed the island landscape and setting, using island resources and bringing together the natural and cultural island heritage.

The Croatian archipelago can be compared only up to a point with other Mediterranean islands in terms of its natural-geographic and socioeconomic features. Despite similarities regarding long-term inhabitation and economic and demographic development, in accordance with the sparse resources of limited space, the Croatian islands are fragmented, and almost all of them are near and rather small. Their fragmentation is relevant geopolitically because they contribute 33,200 km^2 to Croatia's territorial waters. Nevertheless, they sit on the periphery of the Croatian coastal region and the margins of modern social and economic networks.

According to the *Statistical Yearbook of the Republic of Croatia* (Croatian Bureau of Statistics 2018), which lists the inhabited islands, and according to the latest population census recording the island settlements, only 47 of the 1,246 islands, islets, rocks and reefs are permanently inhabited. The islands enjoyed their demographic peak in 1921, when the population numbered 174,994. According to the last population census conducted in 2011, there were only 124,842 islanders, representing 2.8% of the total population of the Republic of Croatia.

Most of these islands have lost their populations rapidly, while the people who remain are ageing. The demographic picture on the bridged islands, those very close to the mainland, and those that are larger is somewhat better, due to better transportation connections and more dynamic economic development, particularly in tourism, fishing, mariculture and (to some extent) olive-growing and wine production. Their economic links are also better, as most of them function as integral parts of the mainland coastal urban region; however, these islands are also exposed to such unfavourable processes as the virtually uncontrolled growth of holiday homes and apartment complexes. The primary economic activities (agriculture, fishing and forestry), which formed the backbone of island development for centuries, have been increasingly abandoned since the mid-20th century, and tourism has become the main focus of economic development. This change has affected the island environment and transformed the Croatian island landscape.

In the past, most island land was used for agriculture. Such landscape patterns as terraces prevailed on the slopes, with drywall structures marking property and preventing soil erosion. Olives and grapes were grown on these terraces. The less fertile parts were used as pastureland and occasionally for forestry. Today, most of these island areas have been abandoned, and indigenous vegetation has taken over. The traditional rural landscape and other aspects of tangible and intangible island cultural heritage are dying out. Diverse interests of the remaining population, visitors to the islands, investors or various policymakers have focused on the narrow belt along the shoreline and the corresponding maritime waters—the area that is considered most attractive nowadays.

Despite the homogeneity of their resources and economic structure, the Croatian islands do not function as a homogeneous region. State policies and legal regulations have not contributed to their greater socioeconomic cohesion. The former nine island and six mainland–island municipalities were abolished in 1992, by new regulations regarding local and regional self-government (OG 90/1992), and the overlarge socialist municipalities were divided into much smaller municipalities and towns commonly regarded as local self-government units (LSUs). Today, the island area is divided between seven mainland counties and as many as 57 LSUs, 8 of which have their seats on the mainland. The administrative-territorial system is as patchy as a leopard's coat. Only 3 LSUs cover individual inhabited islands (Vir, Šolta and Mljet), while 40 cover parts of islands (between 2 and 8 on the same island), and 6 cover larger islands with appertaining smaller islands, where official statistics record at least one settlement (the towns of Mali Lošinj and Komiža, the municipalities of Preko, Sali, Murter-Kornati and Lastovo). As the Pelješac peninsula is treated as an island by the Islands Act, its four municipalities are counted as island LSUs. Other LSUs have seats on the mainland and cover small inhabited islands where there is at least one statistically recorded settlement. For example, the biggest LSU in the Zadar county, the Town of Zadar, administratively covers a range of islands comprising inner, channel and outer subarchipelagos. They are not organised administratively according to their geographical locations and morphostructural parallelism. The island of Pag engirdles this often-confusing division. It is the only island in the EU that is divided between two NUTS 3 regions. The Croatian Bureau of Statistics has reacted rather late to this by as much as two decades. The *Statistical Yearbooks* of the Croatian Bureau of Statistics record small islands with one settlement, where no one has lived for a long time, while islands with sprouting holiday homes, which have permanent residents, are still recorded as uninhabited. To these we should add the islets of Košljun (administratively a part of the town of Krk on the island of Krk) and Galevac (a part of the municipality of Preko on the island of Ugljan), which have been inhabited by monks for centuries.

The existing administrative organisation clearly cannot guarantee synergy of resources or complementarity in economic development. The pragmatics of local particularism has overshadowed the logic of economically justifiable administering the archipelago as a whole. The fragmentation of the island area is further exacerbated by how passenger traffic is organised, with most ferry lines running from the islands to the nearest coastal urban centres, while links between the islands are rare or nonexistent (see Figure 3.1).

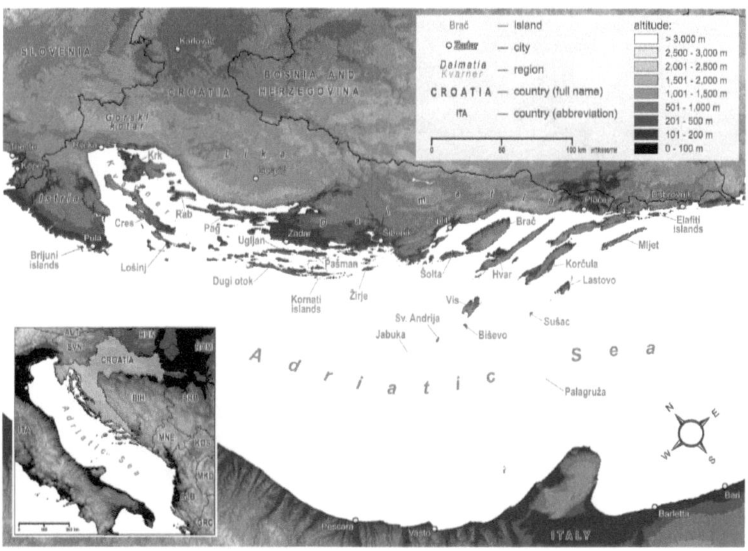

Figure 3.1. Geographic Position and Distribution of the Croatian Islands

POSITION, RANGE, NUMBER AND TYPES

The Croatian islands are located in the eastern Adriatic Sea and comprise 37% of the overall surface area of the Republic of Croatia (with the appertaining territorial waters) (Klemenčić 1992). According to recent research, they cover a total surface area of 3,259.57 km² (Duplančić Leder, Ujević and Čala 2004). In the Croatian geographic literature, it is usual to place them in five groups: western Istrian, Kvarner, north Dalmatian, central Dalmatian and south Dalmatian (Rubić 1952; Stražičić 1987, 1997; Faričić 2012) (see figures 3.2a–3.2c). This division takes into account administrative criteria and nodal-functional regionalisation, and makes sense in terms of natural-geographic and historical-geographic factors (Faričić 2012).

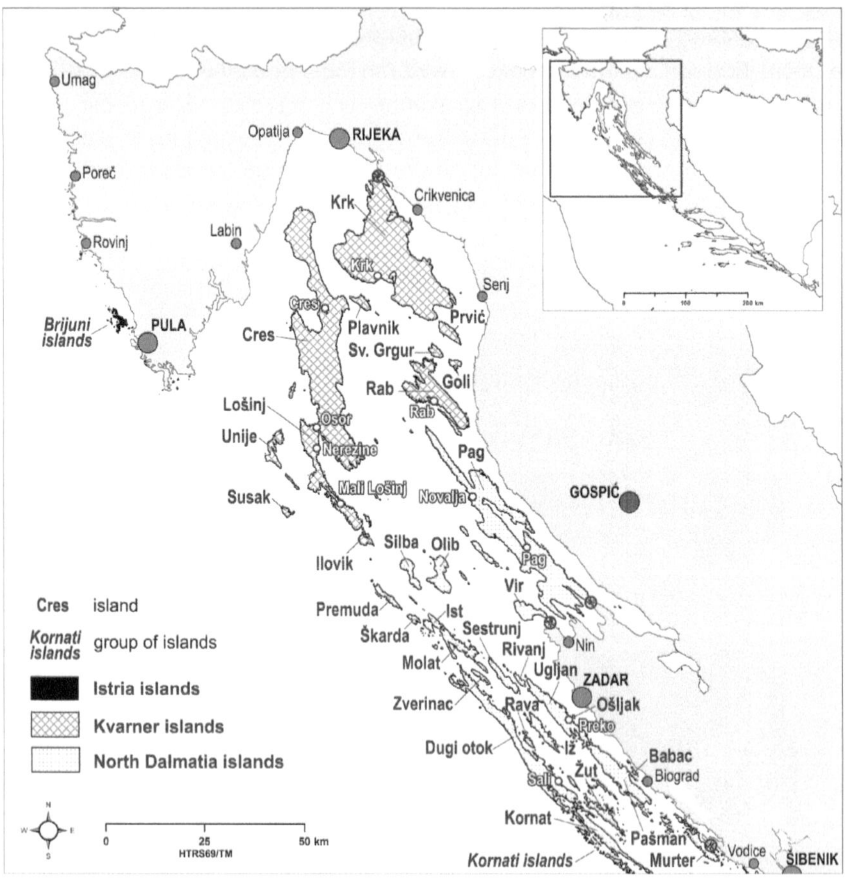

Figure 3.2a.

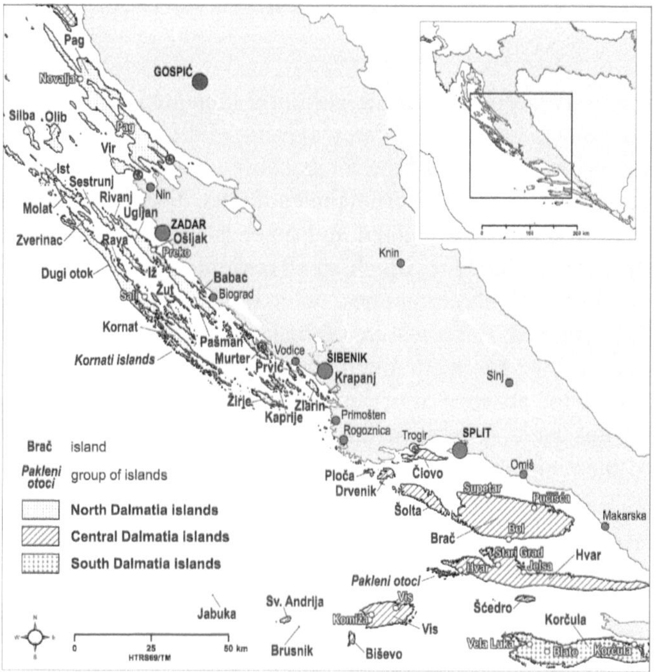

Figure 3.2b.

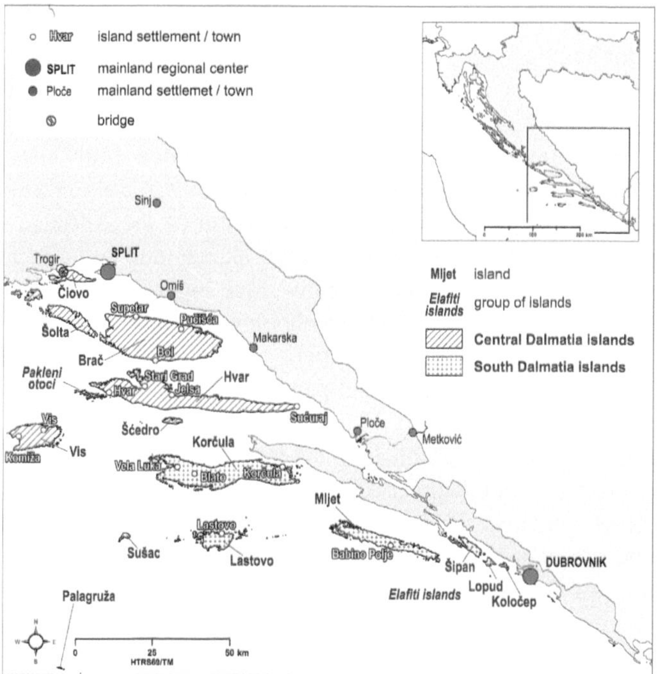

Figure 3.2c. Regionalisation of the Croatian Islands

People have lived on most of the currently inhabited islands more or less continuously since prehistoric times (Čečuk 1996). Throughout history, various factors have influenced the socioeconomic development of particular islands, including their distance from the mainland, dispersion and transportation connections with the mainland and other islands. Towns developed on some of the larger islands to which smaller rural settlements gravitated on the same island or neighbouring ones, while on the smaller islands only rural settlements developed. It is a feature of the north and south Dalmatian islands that connections between them and mainland urban centres were established early on, due to the absence of urban settlements on the islands. Thus, urban subarchipelagos were formed in the two-way communication between the islands and the mainland.

HOW MANY ISLANDS ARE THERE?

The question of the total number of islands is hard to answer and has occupied scientists and experts for many years. Data used in the past, which are still used today as the most recent official information, resulted from analyses of bodies of mainland surrounded by the sea mapped on various types of geographic maps. For centuries, the Croatian archipelago was an essential part of major sailing routes linking Central and Southern Europe with other Mediterranean regions. In an age when there were no standard procedures for selecting geographic content to depict on maps, the depiction of the islands depended on the level of geographic knowledge and the attributes given to them by individual cartographers.

On one of the oldest preserved nautical charts of the Adriatic, made by Pietro Vesconte in 1318, the large Croatian islands and some smaller ones that were essential to navigation are shown (see Figure 3.3). As cartography developed as a science, along with the technology for producing maps, criteria began to develop for depicting geographic objects, so the number of islands depended less on the competence and personal choices of the mapmaker and more on the scale and purpose of the map. Since maps were the primary sources of information for establishing the number of islands and the Croatian archipelago was covered by precise geodetic surveys and maps in increasingly larger scales, the number of islands depicted gradually grew in the literature.

The most recent figure of 1,246 islands, islets, rocks and reefs with the appertaining surface areas (Duplančić Leder, Ujević and Čala 2004) was acquired through surveys using a 1:25,000 scale map, on which rocks and reefs with a surface area of less than 623 m^2 were not usually depicted (on

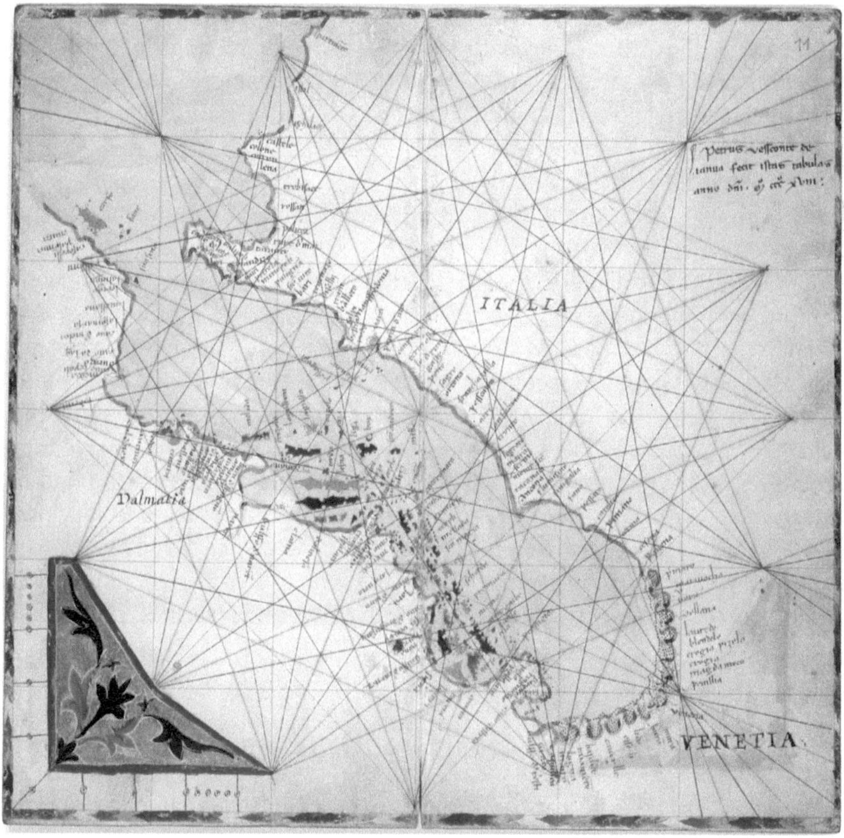

Figure 3.3. The Croatian islands on the Vesconte's portolan chart of the Adriatic Sea, 1318 (Österreichische Nationalbibliothek, Wien, Handschriftensammlung, Cod. 594, Table 11).

the map, they would have been less than 1 x 1 mm), although even objects with a minimum surface area of 25 m^2 could actually have been shown at that scale (0.2 x 0.2 mm on the map). Due to their proximity to the mainland, some islands have changed their status throughout the years and been joined physically to the mainland by infills. This has happened in the old cores of Zadar, Primošten, Dubrovnik and Rovinj, which are mainland cities today (Faričić and Mirošević 2014). On the other hand, some have been separated artificially by dredging channels and abrasion.

Duplančić Leder et al. (2004) did not include some of the smaller bridged islands in their list (for example, Tribunj), although they did include some of the smaller and all of the larger bridged islands (Krk, Pag, Vir, Murter, Čiovo and Trogir). The northern and southern parts of Lošinj were considered distinct islands in research conducted in 2004, because they were separated by the artificial strait of Privlaka (although a road bridge crossing the strait connects them). Privlaka was barely any larger than the bridged Osor Strait, the demarcation point between two indisputably distinct island entities, Cres and Lošinj (Rubić 1952; Stražičić 1975; Lajić and Mišetić 2006). Since this division of Lošinj has not been entirely accepted and a rumour spread that a drop of coffee on one of the topographic maps analysed had been scanned as a reef measuring 625 m², many state institutions in Croatia began to claim that Croatia had 1,244 islands. This was recently incorporated into the Islands Act.

The scientific literature on the Croatian islands reflects (and deepens) the confusion. There are few syntheses dealing with the entire Croatian archipelago, in contrast to the number of comprehensive monographs on individual islands and focussed discussions on narrow archipelago topics in scientific articles. Rubić wrote the only pertinent geographic synthesis (Naši otoci na Jadranu/*Our Islands in the Adriatic*), first published in 1928, and reprinted in 1952 and 2018. Apart from this work, all the Croatian islands have been subjected to the same methodology and dealt with as a consistent unit only in demographic syntheses (Smoljanović, Smoljanović and Nejašmić 1999; Lajić and Mišetić 2006).

ISLANDS AND PSEUDO-ISLANDS

The islands differ from the mainland according to the many implications of physical separation. On the Croatian islands, and assuredly on others, there are many specific characteristics of the plant and animal worlds, and also of sociogeographic development. Many centuries of restricted contact with the mainland and other islands has resulted in islanders relying on limited natural resources, while each island community has developed its own specific speech, customs, culture and way of life, although some features are common to all.

The islandness that emerged from the island way of life is gradually being lost on the Croatian bridged islands. Although they are still surrounded by the sea, they have been integrated into mainland traffic. According to contemporary indicators of spatial, demographic and economic development, they show more similarities with the neighbouring mainland than with the other islands. Some authors who have considered the concept of islands think that the physical connection between an island and the mainland destroys the most important aspect of what makes island life different. For example, Baum (1997) claims that the difference between the mainland and islands has been lost (Royle 2001, according to Baum 1997). This is definitely true of the Croatian islands. Due to their socioeconomic features, they can hardly be classified as "real" islands, so we have called them "pseudo-islands" (Faričić and Mirošević 2014) and their development tendency "functional deinsularisation" (Faričić 2012).

In the latter half of the 20th century, thanks to direct road traffic links with the mainland, tourism developed more rapidly on the Croatian bridged islands than elsewhere in the archipelago. This encouraged the random construction of many holiday homes and apartment complexes, and resulted in significant changes to the island landscape. Some elements of cultural heritage in and around the settlements were jeopardised and often destroyed. This was particularly true of the traditional architecture and landscape patterns associated with a thousand years of agriculture, for instance, terraces, drywalls or stone field shelters. The unbridged islands have also lost some aspects of their traditional landscape, not to unfettered building, but to indigenous vegetation that proliferated after agriculture was abandoned, particularly on the smaller islands and around the smaller settlements. Speaking generally, the rate at which modern people influence the Croatian island landscape rises directly in proportion to the size of the island or settlement and its proximity to the mainland.

LARGE ISLANDS?

The size of an island has a decisive effect on inhabitation, demographic and economic development and islandness overall. So, the Croatian islands should be differentiated in terms of size. International typology that looks at other islands and island states throughout the world is not particularly helpful. For example, Taglioni (2011) has defined small islands as those with surface areas of less than 11,000 km^2 and with fewer than 1.5 million inhabitants. According to these criteria, all the Croatian islands lumped together would still only constitute a single small island with a surface area of 3,259.57 km^2 and a population of about 124,000. Baldacchino (2008) prefers to call such islands "smaller" rather than "small," since most of the world's islands and island states have surface areas of less than 1,000 km^2.

Comparing the size of islands and archipelagos is difficult, because their development has depended on a range of other factors (proximity to other islands or the mainland, the size of the LSU where they belong administratively, the historical-geographic development of the island space and so on). In response to the need to differentiate smaller island groups by typology and better understand specific processes or socioeconomic development patterns, we need to categorise them separately. Several different typologies for the Croatian islands have been produced, taking one or both of the two most common criteria—population and surface area—into account.

Zimmerman (1997) conducted a typisation of the inhabited Croatian islands by surface area. He divided them into four groups: peninsulas and bridged islands, large islands (more than 50 km^2), medium-sized islands (15–50 km^2) and small islands (1–15 km^2); however, he overlooked the fact that there are islands in the Adriatic that are even smaller than 1 km^2, like Ošljak and Krapanj, although these are inhabited and count statistically as individual settlements.

Duplančić Leder et al. (2000) also began with surface areas and distinguished islands (more than 1 km^2) from islets (.01–1 km^2), reefs and rocks (less than .01 km^2). They used the same categorisation when measuring the lengths of island coastlines and surface areas (Duplančić Leder, Ujević and Čala 2004). Lajić and Mišetić (2006) classified the inhabited Croatian islands based on their populations. They divided them into extremely small (less than 50 inhabitants), small (51–250 inhabitants), medium-sized (251–3,000 inhabitants) and large (more than 3,001 inhabitants). This kind of classification is more suited to researching the socioeconomic development of the islands.

Magaš (1993, 2008) calls any Croatian island with three or fewer island settlements "small," regardless of its size or population. Pseudo-islands are excluded from this group because of the application of the utilitarian criterion. At all events, the smaller islands where rural settlements have developed, and that have poor transportation connections with urban centres on the mainland or neighbouring islands, are most at risk in demographic and economic terms. Their socioeconomic development is disabled or hindered, so their populations decrease drastically and sometimes die out altogether (for example, Škarda and Sveti Andrija) (Faričić, Graovac Matassi and Čuka 2010; Faričić 2012). Wherever they are, small islands are hit hardest by spatial isolation, which is mainly reflected in their economic development (Royle 2001). In Croatia, there have been particularly negative consequences for the smallest, remotest islands.

NATURAL RESOURCES: THE BACKBONE OF SOCIOECONOMIC DEVELOPMENT

The Croatian islands are part of the Dinaric karst region, which was mostly formed by deposits of Cretaceous and Eocene sediments, predominantly carbonates (limestone and dolomite), then by folding and fissuring typical of the Alpine orogeny (Herak 1986, 1991). Folds and fissures were formed by geological processes throughout the entire island area, and after the Upper Pleistocene–Holocene rising of the sea level by 121± 5 m (Fairbanks 1989), the islands and their shorelines were formed. In terms of formation, the Croatian islands are coastal (according to Royle's typology, 2001) or ingressional (according to Ratter 2018). The basic lie of the geological structures is NW–SE and is known as the Dinaric line. It is especially pronounced in the area of the north Dalmatian islands. The central and southern Dalmatian islands lie in a W–E direction.

The islands mostly consist of limestone and dolomite, on which thin layers of mostly cambic soils (red and brown) and humus-forming soils (rendzina and limestone-dolomite black soil) have developed alongside climate-vegetation features. A thousand years of exploitation means that they have been anthropogenised and eroded in places (Vidaček, Bogunović, Sraka and Husnjak 1998). There are deeper layers of flysch only on the larger islands situated closer to the mainland, and these can be used for agrarian purposes, for example, on the Kvarner and the north Dalmatian islands. Along with the flysch zones, island plains also play an essential part in inhabitation and the development of agriculture, as they are often filled with a deep layer of red soil. There are also dolomite zones where a thicker pedological cover has formed as the result of the minor mechanical wearing of the stone. Sandy soil is rarer on the Croatian islands (Susak and parts of Unije and Rab) and is mostly used for grape-growing.

Due to the meagre pedological base, the islanders learned rather early on to make the best use of fertile areas, and this resulted in a recognisable island landscape. In the best zones, the island plains, parcelisation was typical. The parcels were rather small, so all the island families could use some of the most fertile lands for vineyards or cereal crops. The slopes were cleared to provide more land, terraced with drywalls to prevent erosion and planted with olives and vines. The harsher karst areas with the least arable land and areas far from settlements were used for grazing. The landscapes formed can be seen on all the inhabited islands and those with no permanent settlements but that were used by islanders, for example, Kornat (see Figure 3.4).

Geological features and processes shaped the island relief. The surface area, innate spatial relations (mainland, channels, islands) and local island erosion bases (shores) were determined on the whole by the rising of the sea level during the Upper Pleistocene and Holocene. Basic geomorphological shapes were created by geological tectonics and dynamics, shaped by fluvio-karst denudation processes, slope processes and abrasion on the island slopes and in the contact belt of the islands and the sea, and accumulation processes in areas of slopes with slight inclines. The recurrent exogenic relief objects on the island are uplands, some in a chain forming the island ridges and some

Figure 3.4. Olive Grove on Kornat, an Agrarian Oasis in a Karst Desert

distinct uplands and lowlands filled with various unbound or semibound sediments (depressions, dry valleys, ravines, boreholes, sink-holes and so on). In relation to their size, the vertical elevation on most islands is relatively high (Vidova Gora, on the island of Brač, is the highest point, at 780 m), although some islands are rather low (Olib, Silba and Susak). Almost all of the Croatian islands have very indented shorelines. Pag is the most extreme example, with a coastline five times longer than the circumference of a circle covering the same surface area. Island coastline indentation includes peninsulas, capes and isthmuses linking the island bays and coves. Settlements have developed in almost all the larger bays and coves.

The island shorelines vary in inclination and height. Structural cliffs have formed along deep fissures (Unije, Dugi Otok, Kornat, Šolta and elsewhere), while abrasion cliffs with abrasion terraces have developed in clastic rock (Vir, Pašman, Vrgada and elsewhere). Many island shores are low, whether formed of rock, pebbles or sand. They are attractive today in terms of tourism.

Both the geological material and mineral composition of the rock are reflected in the appearance of the island settlements, their infrastructure and architectural elements. Stone has long been vital for building houses and outbuildings, drywalls and harbours, and is an integral, essential part of the island anthropogenic landscape. Solid stone blocks of layered limestone are integrated in houses, churches and public buildings in the old island cores. Layered slabs are used to pave the island streets, while irregular large and small stones and fragments provide the necessary material for building drywalls and terraces. Stone has been not only used in construction, but also exported as a natural resource to the neighbouring mainland and farther inland. Brač stone is particularly famous for its appearance and quality, and forms an integral part of the natural and cultural-historical heritage of the island and its identity. Until recently, the village of Pučišće based its economic development on quarrying stone that was cut and dressed locally, then exported as a high-quality island product. The only stonemasonry school in Croatia is still in Pučišće.

In the past, limestone was exploited on many islands to produce lime. Some islands, like Pašman and Škarda in the Zadar group, Žirje in the Šibenik group and Šolta in the Split archipelago, were particularly well known for lime production, and it is even mentioned on some old maps (see Figure 3.5). Lime was used for building houses on the islands, and the mainland towns also required large quantities. Limestone was burned in lime kilns to extract calcium oxide (quicklime) from calcium carbonate and soaked in water to

Figure 3.5. Note on Lime Production on the Island of Žirje, on Kolunić's Map of the Zadar and Šibenik Region, 1570. ***Felbar Collection***

obtain slaked lime. The process required a great deal of wood, and intense tree-felling degraded the island vegetation.

Many archive sources indicate the islanders' obligation to transport certain important island products to mainland towns and reveal the natural resources that some inhabited islands could draw on. At the time when most of the island settlements were formed, and indeed right up to the turn of the 20th century, most island lands belonged to wealthy families from the islands themselves or neighbouring mainland, or ecclesiastical institutions (bishops' households, chapters and monasteries). From the Middle Ages onward, landlords gave the land to the islanders and their heirs to use in perpetuity (this was known as the *colonat* system), in return for tributes that could amount to a third or half of all yields. Through agrarian reform, the islanders began to redeem their island possessions, and the *colonat* system was abolished in the late 19th and early 20th centuries. Islanders, or those with island heritage, then owned most of the island land; however, when mainland landlords still owned the islands, the islanders were often obliged to send part of their olive and grape harvests, and other products, to larger urban centres on the mainland, along with other natural resources. The best example was the Zadar islands, whose inhabitants were required to send timber, stone slabs and lime to Zadar in the 18th century for the maintenance and repair of public buildings.

Natural resources are a salient factor in island development, although some, for instance, fertile soil and mining minerals, no longer play a decisive role in inhabitation or the development of island economies. Water resources are still crucial, whether surface or underground waters. Due to the porous, permeable rock, there are almost no permanent surface watercourses, while ampler sources of water are usually found where the flysch meets the limestone and in dolomite zones. In hydrogeographical terms, Lake Vrana on the island of Cres is a rare, valuable, freshwater phenomenon in the karst. It supplies Cres and Lošinj with water and is considered ecologically and socially to be an exceptionally sensitive and vulnerable environment (Bonacci 2014). Some large islands, for example, Krk, Rab, Pag, Dugi Otok, Vis and Korčula, have somewhat better water resources that can supply most or all of the island settlements. The larger islands close to the mainland have been connected to the mainland water supply systems and no longer rely on natural reserves; however, the small islands and those farther from the mainland have not yet been connected, and during the summer, when demand is higher, they experience water shortages (Gereš 1998). This is a significant barrier to their economic development. Their water supply consists of private reservoirs or tanks that collect rainwater from roofs. Old village wells, which collect rainwater from huge stones and slabs specially laid for that purpose, are also still used. In the summer, water is brought to the small islands by special boats.

The sea, as a natural resource, has had a many-faceted significance in forming the island identity and economic development. It created and developed islandness; on the one hand, it was the medium that separated the islands physically, and on the other, it was a means of communication that facilitated links between settlements on the island coasts, islands or islands and the mainland, when sailing and modern vessels were developed.

The chemical and physical properties of seawater favour biodiversity and various types of economic valorisation. The annual range of Adriatic water temperatures is quite small (from 8°C in winter to 25° in summer, on average). Salinity is 38 per mille, and the water is very clear (Buljan and Zore-Armanda 1976; Cushman-Roisin, Gačić, Poulain and Artegiani 2001). According to various reports of the Ministry of Environmental Protection and Energy of the Republic of Croatia, the water quality is high, even near coastal towns, which are no longer significant pollution sources since the collapse of industry in the 1990s. The only sea factor that affects island life to some extent are high waves raised by strong winds, which make navigation difficult and occasionally impossible (on the open sea, they can reach 10 metres). The island harbours tend to be located in sheltered bays and protected by harbour walls and jetties. The vertical range of sea levels between high and low tides is not vast (measurable in mere tens of centimetres at the conjunction of the Moon and Sun), so it is possible to build harbour infrastructure, housing and other facilities right on the shoreline. The sea in the channels between islands is affected more by tidal currents (particularly in the straits) than by the general direction of the current in the Adriatic (10–20 cm/s), which is anticlockwise (from SE to NE).

There are about 400 species of fish in the Adriatic, of which about 100 are edible (Jardas 1996). Fishing focuses on about 20 species that form island diets and tourist menus. The most common oily fish are sardines, anchovy, mackerel, bonito and tuna, while white fish include various sparidae, for instance, dentex, gilthead, monkfish and bream. There are also bass, hake and mullet, and, more rarely, such high-quality fish as John Dory and grouper. Cephalopods are also regular catches (cuttlefish, squid and octopus), along with crustaceans (shrimp and lobster) and shellfish (mussels and oysters). Mariculture is based on farming fish in cages (bream and bass), catching tuna for cage-feeding, and commercial oyster and mussel beds.

Fishing has always been part of life on the islands as not only an economic activity and source of food, but also a way of life that gave rise to various components of tangible and intangible cultural heritage (wooden fishing boats, the skill of making nets, lobster pots and other fishing tackle, the fishermen's vocabulary and customs and so on). Fishing also encouraged various artistic expressions, from folk singing to high aesthetics in literature, art and music. Coral-fishing was particularly associated with the island of Zlarin, and sponge-diving with the island of Krapanj. Large or small saltworks developed on almost all the inhabited islands, but today only the saltworks on Pag are still active and are still the largest in Croatia. Mariculture has been growing in importance. Although the islands have huge potential, most mariculture companies are registered on the mainland. In 2017, of 158 Croatian companies entered in the Registry of Fish and Shellfish Breeders, only 14 (8.9%) were on the islands. Nonetheless, mariculture has influenced the islands to a considerable degree, although it is a double-edged sword: Fish farms reduce the pressure on wild fish stocks and employ a rather strong contingent of the small island workforce. On the other hand, mariculture installations in sheltered island bays and channels, with slight currents and minimum seawater interchange, harm the marine ecosystem by saturating the organic material.

Shipping, maritime commerce and boatbuilding have contributed to the diversification of island economies. The Croatian islands are part of the Mediterranean maritime system, including particular urban elements in the physiognomy of island settlements, and this has fostered the overall maritime orientation of the Croatian coastal region. In the modern age, seaside and nautical tourism have come to form the core of economic development throughout the island space.

Throughout the centuries, life on the islands has been highly dependent on the climate, its effect on vegetation and agriculture, building and daily and annual patterns of social of economic activities on and around the islands. Today, the climate features of the island are still relevant natural factors in terms of the island economy, which relies largely on tourism based on the sea and weather, with plenty of sunshine and little rain.

The island flora depends very much on the climate. It is dominated by Mediterranean and Southern European types (Kovačić et al. 2008). There are many species, even on the small islands (for example, about 400 plant varieties have been found on Rava, which has a surface area of 3.6 km^2, and 620 on Vir, with a surface area of 22.1 km^2, much of which has been taken over by holiday homes and apartments) (Milović and Pandža 2008, 2016). The most common woody plants are the holly oak, strawberry tree, mastic tree, myrtle and juniper and viburnum, and such shrubs as rock-roses, immortelle, sage and rosemary grow wild, while there are many kinds of orchids in the meadow communities.

There are a wide variety of uses for the island flora, from fuel (mostly oak) and building materials for houses, boats and tools, to foods, dyes and medicinal and aromatic preparations. A large quantity of timber used to be chopped to fire lime kilns and provide fires for night fishing for oily fish, and forests and undergrowth have been destroyed by controlled fires set to clear the land or extend the arable areas and pastures. On many islands, centuries of exploitation have degraded the vegetation, so that forests are now rare (today, those on Mljet, Rab and Pag are under special protection), and the most common plant communities are maquis and garrigue. On the island slopes exposed to wind and sea spray, used for centuries for grazing sheep and goats, only low-growing plants are found, with occasional bare rocky tracts (for example, the slopes exposed to the *bura* on Pag and Rab, and various kinds of degradation of the vegetation cover on the exposed slopes on the Kornati islands and so forth).

During the past few decades, as deagrarianisation has progressed, the process has been reversed, and wild vegetation has taken over, along with the spread of some imported varieties, like the Aleppo pine in the island landscape. Although these forests provide shade near the beaches and have pleasant fragrances that are attractive to tourists, they also reduce the biodiversity and take over former terraced olive groves and vineyards. In the summer, forest fires spread quickly in pine forests and may be hard to extinguish if they are less accessible.

Due to their geographic position at moderate latitudes, the Croatian islands are exposed to secondary air circulation caused by the Azores High in summer and the Icelandic or Genoa Low and Siberian High in winter. Daily air circulation occurs due to differences in the sea and land temperatures, and of course the Adriatic Sea itself, which acts as a modifier of basic climate elements. The Dalmatian islands mostly have the features of Csa or Mediterranean climate with hot summers (according to Koppen's regionalisation), while the Kvarner islands a little further north are in the Cfa climate zone, with a moderately warm, humid climate and hot summers (Šegota and Filipčić 2003). Extreme conditions are rare in this mild climate, but when they do occur, they can be extremely harmful (drought, storms, high winds, frost, hail and so on).

Both climate types have maximum mean monthly air temperatures in July and August of about 24°C and minimum ones in January of about 7°C. Most precipitation falls in the colder part of the year, with only a little during the hot, often arid summer months. When it is most needed, there is little rain, and water reserves on the island are reduced dramatically. The islanders often experience water shortages, even for basic everyday needs, and this, of course, affects the economy. Annual rainfall on the south Dalmatian islands is approximately 800 mm and reaches about 1,000 mm on the Kvarner islands. Snow and other types of precipitation are rare. In the past, when agriculture formed the backbone of development, cultures adaptable to drought were cultivated, particularly olive trees, grapevines, fig, carob, almonds and citrus fruits. To a lesser extent, cultures like tomatoes and green vegetables, which require occasional watering, were grown in gardens or fields with ponds. Today, water shortages (on the islands not connected to the mainland supply) cause problems for tourism development, since most tourists come in the summer when there is less water. Tourism cannot assume mass proportions on the smaller islands, which may be a good thing, as tourists tend to overburden the infrastructure capacity and island space in general. The problem of sewage has still not been systematically addressed on the small islands. There is a lack of modern sewage tanks and filters, and those that exist do not comply with expected ecological standards.

The wind is the climate feature most talked about on the islands. The *bura* (NE), *jugo* (SE) and *maestral* (NW) are the most frequent, strongest winds. During the era of sailing ships, the wind was the crucial natural navigation factor. Today, it remains vital in nautical tourism (Marelić 2016). The wind also clears the air, helps pollination and provides power for wind farms (at the moment, only on the island of Pag); however, it does cause problems when it blows hard in winter. During stormy and hurricane-level *jugo* and *bura* winds, passenger traffic is halted, and the islands are cut off from the mainland.

Large areas of the Croatian islands boast many elements of natural heritage and fall under various forms of protection. Three islands are national parks: Mljet (established in 1960), Kornati (1980) and Brijuni (1983). Two nature parks enjoy a slightly lower level of protection: Telašćica (1988) and the Lastovo archipelago (2006). There are also special reserves, park-woodlands, significant landscapes and natural monuments. They come under the formal natural heritage care system, but the practice has shown (particularly in the national parks) that measures for maintaining the ecosystem are not always balanced against the minimum development requirements of the islanders.

SETTLEMENTS AND POPULATION DENSITY

Many archaeological sites throughout the Croatian islands bear witness to almost constant inhabitation since prehistoric times. It was always based on natural factors. The oldest evidence of human presence on the Adriatic islands has been found in the northwest part of Dugi Otok and dates back to the Paleolithic. The archaeological finds there have been confirmed as being between 40,000 and 60,000 years old (Vujević 2016). There are more Upper Paleolithic finds on most Croatian islands, which were then part of the mainland. In the late Pleistocene and early Holocene, the environment in which human communities were then living changed in many ways. The sea level rose by about 120 metres after the last glacial period, so many parts of the mainland on the Dinaric front became bodies of land surrounded by water. The islands came into existence, and the channels had to be crossed. Seafaring began to develop. In addition, agriculture began to develop and spread from the Neolithic onward (Forenbaher and Miracle 2005). This had a profound effect on the landscape and grew in intensity, with occasional fluctuations, throughout the centuries. In line with changes to their way of life, the people chose specific island locations at different times in history to settle in. During the Paleolithic and Mesolithic, they used open spaces or caves and the surrounding areas, while in the Neolithic, when agriculture was developing intensively, they settled on level agrarian areas (island fields or plains) (Forenbaher and Miracle 2005).

Since there were limited natural resources in rather small areas, networks of island settlements developed according to the position of the best agricultural zones or, if the people engaged in seafaring and fishing, in well-sheltered bays. The island fields, as the central areas of agricultural development, remained the primary locations of settlements right up to the mid-20th century, when primary sector activities began to die out and sailing and tourism transferred the spatial focus of development to the narrow coastlines. The most famous island plain, where stone structures still bear witness to its valorisation from prehistoric times to the present, is Starogradsko Polje Plain on Hvar. It is the largest in the Croatian archipelago. In the 4th century BC, Greeks from Pharos founded the settlement of Stari Grad on Hvar (Čavić 2017; Solarić and Solarić 2009) and partitioned the agricultural land with drywalls. They divided it into 73 rectangular plots measuring 1 × 5 stadia (approximately 185 × 925 m), while some plots were subdivided into smaller parcels. The drywalls and stone huts on the plain have remained untouched for almost 2,500 years and are a well-preserved example of the geometric system of dividing land used in Ancient Greece (Solarić and Solarić 2009). In 2008, the plain was inscribed on the UNESCO world heritage list.

The landscape has changed according to the needs of the people who cultivated the land from age to age and the various social and political systems that succeeded one another on the island. The parcelling of Starogradsko Polje Plain tells a potted history of the ages: the slave-owning system of antiquity, when vineyards and olive groves were planted and crops sown; the early Middle Ages when the field was poorly cultivated, and a good deal of it forested over; the estates of the first Croatian medieval aristocrats, when the mosaic of plough-land and vineyards predominated; the extensive economy of early, poorly developed capitalism; socialist collectivism in the 1950s (Čavić 2017) and the most recent period, when olives have gradually replaced the dominant vineyards and cereal crops, and the plots have become smaller and smaller, subdivided among heirs after the deaths of private owners. Fertile areas on other islands have been used in similar ways.

The Croatian islands developed within the changing fortunes of different states on the eastern Adriatic coast. At the dawn of history, the Histrians, Liburnians, Delmats and other Illyrian communities came into contact with the Greek colonists on Vis, Korčula and Hvar. The Croatian islands became part of a single political and administrative entity for the first time during the era of Ancient Rome. In late antiquity and the early medieval period, the islands were part of Byzantium and later became components of the medieval Croatian kingdom. Venice first took control of many Croatian islands in about 1000 AD and competed for power of the Croatian coastline and islands with the Croatian, and then the Hungaro-Croatian, kingdom up to the early 15th century. From then until the late 18th century, the islands came under the Venetian Republic, apart from the most southerly ones, which were part of the Republic of Dubrovnik. In the late 18th century, the Austrian administration took over. During the short-lived French administration (1805–1813), the Croatian islands became part of the same state for the first time since the Middle Ages, and this continued during the second Austrian administration (1815–1918). It was during the French and Austrian administrations that the Croatian islands encountered modernisation measures for the first time, in the areas of transportation, the economy and education, although they remained in the shadow of state and provincial decision-making centres.

The Croatian island space was again fragmented after the First World War, when the Kvarner islands, Lastovo and Palagruža came under the Kingdom of Italy, while the others became part of the Kingdom of the Serbs, Croats and Slovenes and (beginning in 1929) the Kingdom of Yugoslavia. After the Second World War, the Croatian islands became part of Croatia, which was itself part of Yugoslavia until 1991. These political fragmentations and unifications reflected fragmentations and unifications of nations and states in the turmoiled history of the Balkans and may have guided the research interest away from the archipelago. Hundreds of papers have been written on the history of individual Croatian islands and the history of the states and provinces to which the islands belonged, but there is no comprehensive historiographical synthesis on the history of all the Croatian islands. What we provide here is a brief historical-geographic review based on works published by Faričić (2012) and Magaš (2013).

Whatever the changes of political power and states, the Croatian islands have always been on the margins of the socioeconomic processes on the mainland. Their unfavourable position in relation to the centres of state of regional administration has been mitigated to some extent by their favourable traffic position on the eastern Adriatic sailing routes connecting Central and Southern Europe with Mediterranean regions. Island settlements (for example, Mali Lošinj, Korčula and Orebić on the peninsula of Pelješac), which were on those routes, became part of the Mediterranean seafaring geographical system and gradually became established as sailing and shipbuilding centres.

The network of island settlements developed gradually from classical times to the end of the Middle Ages. They were mostly rural, with the economy based on agriculture and fishing, and it was only on the large islands, where seafaring and trade also played important roles, that towns were founded. The oldest Croatian island towns that date back to antiquity are Krk and Rab. Some ancient towns ceded their urban functions to new, neighbouring urban settlements in the late Middle Ages (for example, Osor gave way to Cres), while Stari Grad on Hvar and the town of Vis on Vis also suffered discontinuity. Some became established as island urban centres during antiquity (Hvar, Korčula, then Pag), while others developed more intensely only in the modern age, primarily due to the development of seafaring and tourism (Mali Lošinj and Novalja on the island of Pag, for example). The small islands developed under the gravitational umbrellas of neighbouring mainland towns, and occasionally neighbouring island towns, and so became permeated by urban functions. There are only a few more recent settlements, mostly founded by administrative separation from older ones at the turn of the 21st century. Up to then, they were regarded as extensions of existing settlements and sprang up through the development of tourism and the construction that accompanied it.

There are about 315 island settlements on the Croatian islands (2011 population census), but only 12 settlements on nine larger islands currently hold the administrative status of towns. Bearing in mind the most frequently used models for differentiating urban settlements in Croatia, according to which administrative status is considered, along with the total number of residents, the proportion of those employed in their place of residence, and proportion of those employed in secondary and tertiary sector activities (Lukić 2012), three settlements could be added to this list. Most urban settlements are on large islands to which smaller, rural settlements gravitate, and this mitigates to some extent the influence of negative demographic processes present on other islands. Island towns cannot be compared to larger urban centres on the mainland if one considers their appearance, economic and demographic development, infrastructure and functions. The largest, Mali Lošinj, had 6,091 inhabitants in 2011, while the smallest, Rab, had only 437 (2011 population census).

Today, the island towns are characterised by slight depopulation and noticeable immigration trends. This is true of not only the Croatian islands, but also places far away, for instance, the Pacific islands (Ratter 2018). There was a marked depopulation process after 1921 (the year in which the demographic peak was reached) throughout the entire archipelago (see Figure 3.6). At the turn of the 20th century, the crisis in grape-growing and wine production, the most important economic sectors in most of the Croatian island space, had a marked effect on depopulation. There were failed or poor harvests, the fertile land had been exploited to the maximum, the price of wine oscillated and vine diseases exacerbated the general crisis in wine production, first Peronospora, then phylloxera, which started spreading in the early 1890s from the Kvarner islands and reached the south Dalmatian islands between 1912 and 1925 (Čuka, Mirošević, Faričić and Graovac Matassi 2017). It utterly ravaged the island vineyards and prompted intense emigration from the islands, mostly overseas to the United States, Chile, Argentina and Australia, swiftly reducing the island populations (see Figure 3.7). Later, people continued to leave, particularly when the Yugoslav borders were opened in the 1960s and the islanders formed part of the Croatian émigré contingent, which departed for more developed European destinations (France, Switzerland, Germany and other countries) and traditional immigrant destinations (the United States, Canada and Australia). Industrialisation and urbanisation on the mainland had a further effect on island depopulation, particularly in the north and south Dalmatian island groups, and also on the small islands in other groups; however, depopulation is a feature of other islands throughout the world, particularly those with scarce natural resources, services and functions, which tend to have a weaker social life (Royle 2001).

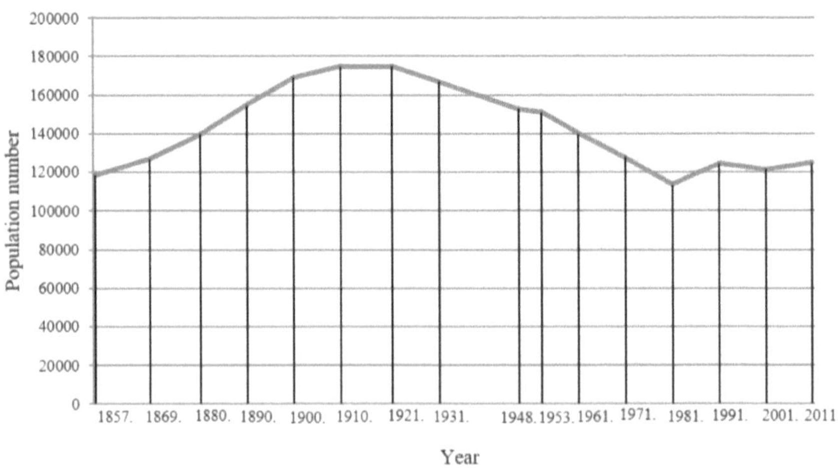

Figure 3.6. Trends in Population Numbers on the Croatian Islands, 1857–2011

Trends among the population numbers of larger urban or urbanised centres can be traced in settlements with the administrative status of towns in relation to the overall numbers of islanders in the same periods (see Figures 3.6 and 3.7) and are important for the demographic development of the island space. Up until the second half of the 20th century, even in urban island settlements, the backbone of economic development was agriculture, particularly grape-growing, and these places also experienced their demographic peak before the onslaught of phylloxera. If we consider overall trends in the populations of the 12 administrative towns, the demographic peak was recorded in 1910; however, while the entire island space lost inhabitants in later periods, the island towns made a demographic recovery after the Second World War and between 1961 and 1981, when there was a huge demographic decrease in the islands overall. They continued to grow up to the most recent population census in 2011. The reason for this can be traced to the economic development of the towns, particularly the development of tourism and other infrastructure that were much more advanced than in the rural settlements. The island towns have direct transport links with the mainland (bridges and roads) or frequent ferry services, their own water supplies or connections to the mainland system and primary (sometimes secondary) schools, doctors and dentists and so on. Apart from tourism, they rely on the development of other economic branches, for inatance, other service industries, along with saltworks (Pag), fishing (Komiža and Vis), seafaring (Mali Lošinj), trade and so on.

Regarding the island population density according to types of settlements in 2011, 33.9% of the islanders lived in 15 urban settlements (12 administrative towns and three urban settlements, according to the aforementioned

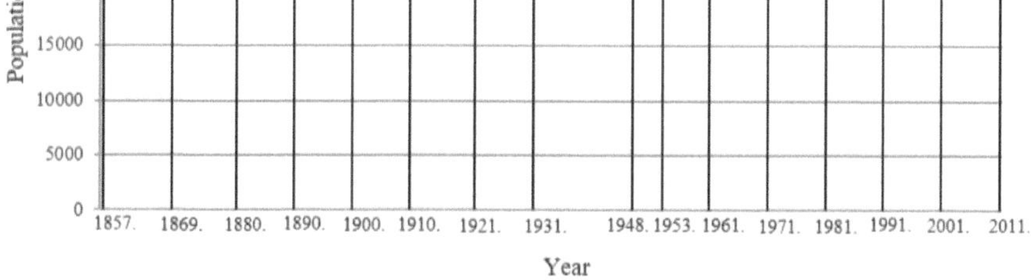

Figure 3.7. Population Trends in the 12 Island Settlements with the Status of Towns, 1857–2011

classification), so it can be concluded that the Croatian archipelago is still mostly rural, with a rather low degree of urbanisation. It is no wonder that the mainland towns and cities have attracted many islanders, and caused negative demographic and economic trends on most islands.

Regardless of the falling number of islanders and the fact that the populations of some settlements and entire islands are on the verge of dying out (Lajić and Mišetic 2013), the number of dwellings is steadily rising thanks to tourism. In 1971, 80.3% of the homes on the islands were used as primary residences (see Table 3.1). Between 1971 and 2011, the overall number of islanders fell by 2.3%, but the number of dwellings rose 3.4 times. Paradoxically, the greatest increase in the number of dwellings (4.4 times) was recorded in the north Dalmatian island group, which lost the greatest number of residents during the same period. As much as 68.1% of dwellings there were reserved for holidays, recreation or tourist rentals, and only 21.1% were used as permanent residences. It turns out that the majority of dwellings on all islands in 2011 (58.5%) were used for holidays. The Croatian islands are being transformed from a working-residential space to one given over to holidays and recreation. This trend is particularly evident on the small islands (Faričić, Graovac Matassi and Čuka 2010).

These processes have changed the physiognomy of the island settlements. In the first place, many historical cores farther from the shoreline have been wholly or mostly deserted. As seafaring and tourism became more important, the inhabitants of inland villages moved down the slopes to the bays and other attractive areas on the shorelines. In places where there has been intense construction in the last decades, the buildings are not grouped around a square and parish church (or cathedral, on the larger islands) in the traditional way but have spread spontaneously along the narrow coastal belt, which is the most attractive area. Some, like Gajac and Mandre on the island of Pag, are new. These new zones, which are mostly residential, are often not equipped with suitable communal infrastructure, lack water supplies and sewage facilities and have poor roads. They lack shopping facilities for food and general goods and have few restaurants or other catering facilities. Haphazard building has neglected to account for contemporary urban standards. The appearance of this new residential architecture is often shoddy, and the new building materials used (concrete and aluminium plastic) have strange shapes, colours and dimensions that do not fit with the traditional island architecture and building patterns. They cannot be claimed as examples of postmodern architecture based on social theories but are the results of quickly produced designs tailored to the pockets of the investors. As a result, they are incredibly diverse in terms of (the lack of) aesthetics and function.

Table 3.1. Use of Dwellings on the Croatian Islands, 1971–2011, by Island Group

			Dwellings for Permanent Residence				*Dwellings for Occasional Residence*		*Dwellings Used for Activities Only*	
ISLAND GROUP	*Year*	*Number of Dwellings*	*Total*	*Inhabited*	*Sometimes Inhabited*	*Abandoned*	*Holidays and Recreation*	*Seasonal Activities*	*Tourist Rentals*	*Other Activities*
Kvarner	1971	15,304	11,525	9,916	255	1,354	3,633	146	—	—
	2011	51,454	20,049	15,238	4,281	530	22,415	14	8,875	101
North	1971	13,482	11,015	10,062	207	746	2,008	459	—	—
Dalmatia	2011	59,466	18,358	12,569	5,024	764	32,513	568	7,973	39
Central	1971	13,121	10,534	9,882	166	486	2,197	390	—	—
Dalmatia	2011	45,056	21,731	14,150	6,693	888	15,533	236	7,488	58
South	1971	7,745	6,815	6,163	88	564	842	88	—	—
Dalmatia	2011	14,637	9,444	6,484	2,175	785	2,845	117	2,207	17
TOTAL	1971	49,652	39,889	36,023	716	3,150	8,680	1,083	—	—
	2011	170,613	65,582	48,441	18,173	2,967	73,306	935	26,543	215

Source: Popis stanovništva i stanova 1971., Stanovi – korišćenje i nastanjena lica, rezultati po naseljima i opštinama, Knjiga I, Savezni zavod za statistiku, Beograd, 1972; Stanovi prema načinu korištenja, Tablica 3.1.1, Popis 2011., Državni zavod za statistiku Republike Hrvatske.

The extent of this kind of tourist colonisation would be even greater if it were not for the fact that building costs are considerably higher on the islands than on the mainland (due to the costs of transporting building materials from the mainland and the lack of qualified construction workers and craftsmen). Another hindrance is the legal-proprietary aspect. Since the islands, particularly the rural parts, were neglected for decades during the time of the Kingdom of Yugoslavia and socialist Yugoslavia, the cadastral records and court-registered land books do not correspond to the actual situation regarding most island parcels.

THE BLESSING AND CURSE OF TOURISM

Since the 1970s, the most important economic activity in the Croatian island space has been tourism. Before then, when such primary sector activities as agriculture, livestock farming and fishing were the principal sources of employment for the active islanders, most of them were thus engaged. Today, tertiary activities dominate the island economy.

Changes in the use of agricultural land between 1960 and 2011 show that agricultural activities have decreased considerably and that the island landscape patterns have changed. No less than 24 islands had to be excluded from our analysis, as there were no separate data for them in 2011. From the perspective of the archipelago with only 47 inhabited islands, the list appears too long: Unije, Susak, Vele Srakane, Ilovik, Olib, Silba, Premuda, Ist, Molat, Rava, Iž, Vrgada, Kaprije, Krapanj, Žirje, Prvić, Zlarin, Drvenik (Drvenik Veli), Ploča (Drevnik Mali), part of Čiovo, Šipan, Lopud and Koločep, each one inhabited but getting older and older, gradually losing strength to preserve their islandness.

Treated as suburbs of mainland cities or somewhat remote parts of mainland and island towns and municipalities, they were statistically aggregated at the level of the LSUs to which they belonged; however, the agriculture areas on these islands comprised only 8.3% of the overall island agricultural land in 1960, so omitting them is not statistically relevant. According to the 1960 agricultural census data, 136,279 hectares were being cultivated (on all the islands). In 2011, the figure was 22,536 hectares, or a reduction of 83.5%. The area of plough-land decreased in the same period from 8,136 to 926 hectares (–88.6%). Livestock farming suffered especially and was only present to a higher degree on Pag. The number of sheep on Pag rose by 22.3%, and Pag sheep's cheese and lamb became well-known, branded island products that today fare well on the international market and win accolades at world food fairs.

Abandoning livestock farming on the other islands has had a significant effect on the landscape, and former pastureland has reverted to indigenous vegetation (holly oak, strawberry trees, mastic trees and myrtle) and allochthonous (Aleppo pine) species (see Figure 3.8). While the decline in agriculture has resulted in unprecedented transformations of the agricultural landscape, the haphazard, uncontrolled development of tourism has transformed the island coastal zones, particularly the most beautiful bays, where the natural karst has been covered in concrete, and vineyards and olive groves

Figure 3.8a. Landscape Changes on the island of Ist. The Situation in the Early 20th Century. ***Photograph from the collection of Zvonko Šuljak.***

Figure 3.8b. The Same Location Photographed in 2010

bulldozed to make room for rental accommodations. Large hotel complexes and holiday villages have sprung up on the large islands. Most tourist activities on the small islands are based on private accommodation dominated by self-catering apartments and rooms.

In the last decades, tourism has boomed in Croatia. The coastal area is the leading destination; most tourist activities take place on the mainland coast and islands. According to numerous reports of the Ministry of Tourism in the last decade (2009–2017), tourist visits to the Croatian islands have increased by 22.2%, and overnight stays by 27%. In 2017, 3.2 million tourists, mostly foreign, visited the islands, and the number of overnight stays was 20.7 million. Given the total number of islanders on all the islands (about 125,000) and the fact that the vast majority of tourists come in June, July and August, it is not hard to imagine the enormous pressures imposed on the weak island infrastructure. Added to this, 48% of tourist visits and overnight stays occurred on just five bridged islands. Here is the dichotomy; tourism is a strategic sector of the economy, but consistent development policies are hardly ever implemented, despite the existing legal regulations. Little is invested in improving the infrastructure, particularly on the more remote unbridged islands, while the islands garner 18.4% of tourist visits and 24.1% of overnight stays in Croatia.

The socioeconomic situation on the small islands is particularly bad. Of 47 inhabited islands, 28 are not bridged and have as many as three settlements, so they belong to the small inhabited island category (Magaš 1993, 2008). In 2011, of all the islanders, only 4,568 (3.7%) lived on these islands. The population older than age 15 was mostly inactive (72.3%) and consisted mostly of retired people (67.5%). Only 27.7% of the population older than age 15 was active, which must raise questions about any serious attempts to revitalise these islands. In contrast, on all the islands taken together, 53.2% were inactive, and 66.8% of the population older than age 15 were active. Among the inactive population, 59.3% were retired people.

The proportion of unemployed people on the islands in 2011 was hard to establish accurately, although analysis was conducted in demographically and economically vital settlements with populations of more than 500 age 15 and older. There were 54 such settlements, where 72.6% of all islanders of that age resided. The proportion of those unemployed was 6.4%, but this low rate is not necessarily due to favourable employment opportunities, but instead emigration, notably by younger, well-educated people who cannot find jobs in their respective professions on the islands and often decide to leave. The high proportion of senior citizens generates many issues regarding social welfare and health care. Due to the lack of highly qualified personnel to mitigate these issues to some extent, the state is sometimes unwilling to take significant steps to ensure the minimum conditions for a dignified old age, let alone the long-term biological revitalisation of the island community.

Elderly populations dominate on the small inhabited islands and in small, more remote inland villages in the interiors of large islands. It seems likely that soon, many small islands and small island settlements will be transformed from permanent residences to resorts only inhabited during the tourist season. The larger islands, particularly those with better connections to the mainland, and those that are functionally closer to it, will continue to enjoy the prospect of being dynamic elements in the island–mainland socioeconomic systems.

REFERENCES

Baldacchino, Godfrey. 2008. "Studying Islands: On Whose Terms? Some Epistemological and Methodological Challenges to the Pursuit of Island Studies." *Island Studies Journal* 3, no. 1: 37–56.

Batović, Šime. 1974. "Prapovijesni ostaci na zadarskom otočju." *Diadora* 6: 5–165.

Bonacci, Ognjen. 2014. "Analiza varijacija razine vode jezera Vrana na otoku Cresu." *Hrvatske vode* 22: 337–46.

Buljan, Miljenko, and Mira Zore-Armanda. 1976. "Oceanographical Properties of the Adriatic Sea." *Oceanography and Marine Biology: An Annual Review* 14: 11–98.

Čavić, Aldo. 2017. "Starogradsko polje kao kultivirani poljoprivredni krajolik kroz povijest." *Prilozi povijesti otoka Hvara* 13: 29–43.

Čečuk, Božidar. 1996. "Špilja Kopačina kod Donjeg Humca na otoku Braču." *Arheološki radovi i rasprave* 12: 13–23.

Croatian Bureau of Statistics. 2018. *Statistical Yearbook of the Republic of Croatia*. Zagreb: Prosinac.

Čuka, Anica, Lena Mirošević, Josip Faričić, and Vera Graovac Matassi. 2017. "Phylloxera Revisited: The Spread of Grapevine Disease in Dalmatia and Its Influence on Socioeconomic Development and Agricultural Landscape." *Annales, Series Historia et Sociologia* 27, no. 1: 101–18. https://doi.org/10.19233/ASHS.2017.09.

Cushman-Roisin, Benoit, Miroslav Gačić, Pierre-Marie Poulain, and Antonio Artegiani. 2001. *Physical Oceanography of the Adriatic Sea*. Dordrecht, Boston, and London: Kluwer Academic.

Državni zavod za statistiku. "Naselja i stanovništvo Republike Hrvatske 1857–2001." https://www.dzs.hr/hrv/DBHomepages/Naselja%20i%20stanovnistvo%20Republike%20Hrvatske/Naselja%20i%20stanovnistvo%20Republike%20Hrvatske.htm (accessed 5 June 2020).

Državni zavod za statistiku. 2017. "Popis stanovništva 2011. Stanovi prema načinu korištenja, Tablica 3.1.1." https://www.dzs.hr/Hrv_Eng/publication/2016/SI-1586.pdf (accessed 5 June 2020).

Državni zavod za statistiku. 2013. "Popis stanovništva 2011. Stanovništvo prema starosti i spolu po naseljima." https://www.dzs.hr/Hrv_Eng/publication/2012/SI-1468.pdf (accessed 5 June 2020).

Duplančić Leder, Tea, Tin Ujević, and Mendi Čala. 2004. "Coastline Lengths and Areas of Islands in the Croatian Part of the Adriatic Sea Determined from the Topographic Maps at the Scale of 1:25,000." *Geoadria* 9, no.1: 5–32. https://doi.org/10.15291/geoadria.127.

Duplančić Leder, Tea, Tin Ujević, Mendi Čala, and Ivana Viđak. 2000. "Categorization and Number of Islands in the Republic of Croatia." *Periodicum Biologorum* 102, no. 1: 281–84.

Fairbanks, Richard G. 1989. "A 17,000-Year Glacio-eustatic Sea Level Record: Influence of Glacial Melting Rates on the Younger Dryas Event and Deep-Ocean Circulation." *Nature* 342: 637–42. https://doi.org/10.1038/342637a0.

Faričić, Josip. 2012. *Geografija sjevernodalmatinskih otoka*. Zagreb: Školska knjiga.

Faričić, Josip, and Lena Mirošević. 2014. "Artificial Peninsulas and Pseudo-Islands of Croatia." *Annales: Anali za Istrske in Mediteranske studije. Series historia et sociologia 24, no.* 1: 113–28.

Faričić, Josip, Vera Graovac Matassi, and Anica Čuka. 2010. "Mali hrvatski otoci—radno-rezidencijalni prostor i/ili prostor odmora i rekreacije." *Geoadria* 15, no. 1: 145–85. https://doi.org/10.15291/geoadria.548.

Forenbaher, Stašo. 2002. "Prehistoric Populations of the Island of Hvar: An Overview of Archaeological Evidence." *Collegium Antropologicum* 26, no. 1: 361–78.

Forenbaher, Stašo, and Preston Miracle. 2005. "The Spread of Farming in the Eastern Adriatic." *Antiquity* 79, no. 305: 514–28. https://doi.org/10.1017/S0003598X00114474.

Gereš, Dragutin. 1998. "Gospodarenje vodom na otocima i vodoopskrba." In *Voda na hrvatskim otocima*, edited by Dušan Trninić, 25–44. Zagreb: Hrvatsko hidrološko društvo.

Herak, Milan. 1991. "Dinaridi – mobilistički osvrt na genezu i strukturu." *Acta geologica* 21, no. 2: 1–83.

Herak, Milan. 1986. "A New Concept of Geotectonics of the Dinarides." *Acta geologica* 16, no. 1: 1–42.

Jardas, Ivan. 1996. *Jadranska ihtiofauna*. Zagreb: Školska knjiga.

Klemenčić, Mladen. 1992. "Geografski smještaj i položaj Hrvatske." *Geografski horizont* 38, no. 2: 7–15.

Kovačić, Sanja, Toni Nikolić, Mirko Ruščić, Milenko Milović, Vanja Stamenković, Darko Mihelj, Nenad Jasprica, Sandro Bogdanović, and Jasenka Topić. 2008. *Flora jadranske obale i otoka—250 najčešćih vrsta*. Zagreb: Prirodoslovno-matematički fakultet and Školska knjiga.

Kozličić, Mithad. 1995. *Kartografski spomenici hrvatskog Jadrana. Monumenta cartographica Maris Adriatici Croatici*. Zagreb: AGM.

Lajić, Ivan, and Roko Mišetić. 2013. "Demografske promjene na hrvatskim otocima na početku 21. stoljeća." *Migracijske i etničke teme* 29, no. 2: 169–99. https://doi.org/10.11567/met.29.2.3.

Lajić, Ivan, and Roko Mišetić. 2006. *Otočni logaritam – Aktualno stanje i suvremeni demografski procesi na jadranskim otocima*. Zagreb: Institut za migracije i narodnosti i Ministarstvo mora, turizma, prometa i razvitka.

Lukić, Aleksandar. 2012. *Mozaik izvan grada, tipologija ruralnih i urbaniziranih naselja Hrvatske*. Samobor: Meridijani.

Magaš, Damir. 2013. *Geografija Hrvatske*. Zadar: Sveučilište u Zadru, Odjel za geografiju and Izdavačka kuća Meridijani.

Magaš, Damir. 2008. "Geografske posebnosti razvitka malih hrvatskih otoka." In *Otok Rava*, edited by Josip Faričić, 19–42. Zadar: Sveučilište u Zadru, Razred za prirodne znanosti HAZU, Matica hrvatska—Zadar and Hrvatsko geografsko društvo – Zadar.

Magaš, Damir. 1993. "Prometnice malih jadranskih otoka. " *Geografski glasnik* 55: 151–55.

Marelić, Tome. 2016. "Wind Influence on Sailing Ship Navigation across the Croatian Part of Adriatic Sea." *Geoadria* 21, no. 2: 211–36. https://doi.org/10.15291 /geoadria.21.

Milović, Milenko, and Marija Pandža. 2016. "Flora i vegetacija otoka Vira." In *Otok Vir*, edited by Damir Magaš, 97–130. Zadar: Sveučilište u Zadru and Općina Vir.

Milović, Milenko, and Marija Pandža. 2008. "Flora otoka Rave te okolnih otočića i hridi." In *Otok Rava*, edited by Josip Faričić, 225–64. Zadar: Sveučilište u Zadru, Razred za prirodne znanosti HAZU, Matica hrvatska – Zadar, Hrvatsko geografsko društvo – Zadar.

Ministarstvo zaštite okoliša i energetike. 2017. *Nacionalno izvješće o godišnjoj i konačnoj ocjeni kakvoće mora na plažama hrvatskog Jadrana u 2017. godini* (National Report on Annual and Final Evaluations of Sea Quality on Croatian Adriatic Beaches, 2017). Zagreb: Ministarstvo zaštite okoliša i energetike.

Ratter, Beate M. W. 2018. *Geography of Small Islands: Outposts of Globalisation*. Cham: Springer International.

Royle, Stephen A. 2001. *A Geography of Islands: Small Island Insularity*. Routledge Studies in Human Geography. London and New York: Routledge.

Rubić, Ivo. 1952. *Naši otoci na Jadranu*. Split: Izdanje Odbora za proslavu desetgodišnjice Mornarice.

Šegota, Tomislav, and Anita Filipčić. 2003. "Köppenova podjela klima i hrvatsko nazivlje." *Geoadria* 8, no. 1: 17–37. https://doi.org/10.15291/geoadria.93.

Smoljanović, Mladen, Ankica Smoljanović, and Ivo Nejašmić. 1999. *Stanovništvo hrvatskih otoka*. Split: Zavod za javno zdravstvo Županije splitsko-dalmatinske.

Solarić, Miljenko, and Nikola Solarić. 2009. "Laying Out Land Parcels and the Oldest Boundary Stone in Croatia from the 4th Century BC." *Cartography and Geoinformation* 8, no. 12: 59–77.

Stražičić, Nikola. 1997. "Prilog poznavanju ukupnog broja hrvatskih jadranskih otoka i broja naseljenih otoka među njima." *Pomorski zbornik* 35: 219–40.

Stražičić, Nikola. 1987. "Prirodno-geografske značajke kao poticajni i ograničavajući faktori razvoja jadranskih otoka." *Pomorski zbornik* 25: 39–55.

Stražičić, Nikola. 1975. "Cresko-lošinjska otočna skupina." *Otočki ljetopis Cres-Lošinj* 2: 143–90.

Taglioni, François. 2011. "Insularity, Political Status and Small Insular Spaces." *International Journal of Research into Island Cultures* 5, no. 2: 45–67.

Vidaček, Željko, Matko Bogunović, Mario Sraka, and Stjepan Husnjak. 1998. "Stanje i mogućnosti korištenja zemljišta jadranskih otoka." In *Voda na hrvatskim otocima*, edited by Dušan Trninić, 163–77. Zagreb: Hrvatsko hidrološko društvo.

Vujević, Dario. 2016. "Adriatic Connections: Exploring Relationships from the Middle Palaeolithic to the Mesolithic." In *Croatia at the Crossroads: A Consideration of Archaeological and Historical Connectivity*, edited by David Davison, Vincent Gaffney, Preston Miracle, and Joanna Sofaer, 19–32. Oxford, UK: Archaeopress.

Zimmerman, R. 1997. "Strategija razvitka otoka s prostorno-planskih gledišta." In *Nacionalni program razvitka otoka*, 33–126. Zagreb: Ministarstvo razvitka i obnove.

SOURCES

Savezni zavod za statistiku. Beograd. 1972. "Popis stanovništva i stanova 1971. Stanovi–korišćenje i nastanjena lica. Rezultati po naseljima i opštinama. Knjiga I.

HR-DAZD-359, *Fond obitelji Lantana*, box 1, call no. 43, anagraphic table from 1759, Zadar State Archives.

HR-DAZD-359, *Fond obitelji Lantana*, box 2, call no. 45, label 8, anagraphic table from 1759, Zadar State Archives.

Österreichische Nationalbibliothek, Vienna, Handschriftensammlung, Cod. 594, Table 11.

Državni zavod za statistiku Republike Hrvatske. 2011. Popis stanovništva, kućanstava i stanova 2011. Stanovi prema načinu korištenja, Tablica 3.1.1.

Chapter Four

The Demography of the Archipelago

Migration as a Way of Life

Sanja Klempić Bogadi and Sonja Podgorelec

Ista sudba
S otoka na otok
ne možeš skočiti
ražet forcon voge
moraš prebroditi.
Ka baciš barbetu
i čapaš *se kraja*
vidiš da bodule
ista sudba spaja.

The same fate
From island to island
you cannot jump
but best the wind
with oar power.
When you cast a rope
and reach land
you see the islanders
share the same fate.

—Toni Cukrov, 2009, lines written in
the idiom of the island of Prvić

Near islands, like those in Croatia, give rise to specific issues, primarily about the people who live on them. Who forms the modern island communities in the Croatian archipelago, and how do they live? What affects their numbers, which change so dramatically from time to time? What are the dominant features of everyday life in the island settlements? What are the basic life values on the islands, and how do the inhabitants assess them?

Answers can be found by reviewing the basic demographic and socio-economic processes affecting the size, structure and way of life of Croatian island communities. Their development, stagnation or reduction are partly determined by their geographic position and distance from the mainland centres to which they gravitate. The Croatian archipelago is classed as "near islands," but Croatian scientists tend to take an Adriatic view rather than the Mediterranean view and structure it according to the distance from the mainland. When analysing demographic, economic and overall social development, they highlight the difference between coastal and outer islands (Nejašmić 1991, 2013) or coastal, channel and open-sea islands (Klempić and Lajić 2005). As for the concept of islandness, the proximity of the mainland has always been a significant factor.

This chapter is based on an analysis of census data and information gathered throughout the course of several research projects. Since migration influences change in demographic structures, along with various aspects of the islanders' way of life, it must be examined during a longer period of time. Here, we present the results of six field research projects carried out between 1990 and 2013, on islands varying in size, number of inhabitants and distance from the mainland. We carried out our investigations on coastal islands close to the mainland (Krapanj, Krk, Prvić, Zlarin, Drvenik Veli, Ugljan and Šolta), those a little farther out or located in channels (Cres, Iž, Kaprije and Lošinj) and distant, outer islands (Žirje, Silba and Dugi Otok). Along with firsthand observations and questionnaires, we also used semistructured interviews, that is, a biographical approach to collecting information about the way and quality of life of the islanders and the role migration played during various phases of their lives, whether their own or their parents'. The subjects of the research were inhabitants belonging to various age groups. To convey the atmosphere of island life as closely as possible and support our findings, some of the respondents' statements are given (their answers to questions or some parts of their life stories).

MIGRATION AS A WAY OF LIFE

Continual migration triggered by various outside influences is a strong counterweight to the isolation that defines the daily lives of people living in the limited space of an island with limited natural resources and social capital. From the time of the first settlements, the Adriatic islands were subject to different kinds of migration. Until the mid-19th century, most movements occurred from the mainland to the islands, followed by waves of migration from the islands to the mainland. Today, the islands are perceived as areas of emigration and depopulation; however, as noted in the previous chapter, population censuses have recorded growth in the number of islanders from the early 1990s to the present day.

Population Growth: 1857–1921

The first modern population census was carried out in 1857, and recorded 118,107 inhabitants on the islands (see Table 4.1). Economic prosperity was primarily due to grape-growing, wine production and sales, which influenced social relations and improved living standards in island households. Planting new vineyards and increasing wine production resulted in higher household incomes, and after long centuries of the tenant–landlord system known as *colonat*, proprietorial relationships began to change. Tenants were able to redeem their land from their landlords, and this had a positive effect on demographic processes. The number of islanders rose constantly, mostly as a result of the high natural growth rate, but also because of immigration (Lajić 1992). The highest growth was reached between 1880 and 1890, after which it slowed because of emigration. Agricultural crises prompted strong emigration waves. The Wine Clause of the 1891 trade agreement between the Austro-Hungarian Empire and Italy, which permitted Italian wines to be sold in Croatia and the entire Monarchy, and the vine pest phylloxera, caused economic hardships of catastrophic proportions. By then, the island economies had more or less switched from diversity to monoculture and relied almost exclusively on grape-growing and wine production. The market for Croatian island wines, however, was lost.

The collapse of viticulture as the basic economic activity meant many people were without work and could not find jobs in other, undeveloped branches of the island economy. The main factors that prompted emigration were agrarian overcrowding, mounting debts in relation to redeeming land and the declining use of the sailboats on which most islanders relied (Lajić 1989; Nejašmić 1991; Podgorelec and Klempić Bogadi 2013).

The trickle of emigration in the late 19th century swelled into a flood in the early 20th century, with people departing mostly to the United States, Chile, Argentina, Australia and New Zealand. It left a permanent mark on the demographic and socioeconomic development of the islands; however, the strong natural growth rate, mostly due to the high birth rate, made up for losses (emigrants and those killed in the First World War), and in the early decades of the 20th century the total island population grew. In 1921, the highest number ever was recorded—174,994.

Depopulation, which was partly masked by the 1921 island population data, had begun on the larger Dalmatian islands (Brač, Hvar and Šolta) early in the 20th century, mostly due to increasing emigration abroad. The small islands only began to significantly lose their populations just before the First World War (Lajić 1992). News of good wages to be made abroad prompted the islanders, mostly the men, to migrate temporarily, although this often became permanent. Their womenfolk stayed behind with their children and elderly relatives, and ran the households, but also worked in the gardens and fields, and fished (traditional men's jobs). This continued after the Second World War. An older person we spoke to recalled the following:

> The wife would be left on her own at home and worked while the husband sailed, and she was on Kaprije. She did everything, you see. We used to do everything, dig as well. The women went fishing, more than the men, much more. —female, aged 91, Kaprije, 2011

The sheer number of young men who left, regardless of their marital status, had a huge effect on the formation of new families and fertility on the islands. Later, young women resorted to moving away from the islands to marry non-islanders (Podgorelec and Klempić Bogadi 2013).

Table 4.1. Trends in Population Numbers on the Croatian Islands, 1857–2011

1857	*1869*	*1880*	*1890*	*1900*	*1910*	*1921*	*1931*
118,378	128,992	139,870	154,837	168,240	174,579	174,994	167,074
1948	*1953*	*1961*	*1971*	*1981*	*1991*	*2001*	*2011*
152,962	151,060	140,218	127,814	113,299	124,296	122,418	124,955

Source: Author's calculations according to data from *Settlements and Population of the Republic of Croatia, 1857–2011*, Croatian Bureau of Statistics, Zagreb, CD ROM; *Census of Population, Households and Dwellings 2001*, Croatian Bureau of Statistics, Zagreb, CD ROM; *Census of Population, Households and Dwellings 2011*, Croatian Bureau of Statistics, Zagreb, www.dzs.hr.

Depopulation: 1921–1991

During the seven decades or so of that period, war and various crises struck the island economy; fishing and fish products, olives, figs, vegetables, small livestock, boatbuilding and seafaring were affected. Between 1921 and 1948, the island population dwindled due to emigration caused by economic crises and the loss of lives in the Second World War. Emigration continued, but now, in addition to heading for destinations across the oceans, the islanders increasingly chose Western European countries and cities on the Croatian coast. Our respondents recalled how the island communities on the Kvarner and Dalmatian islands had decreased in the last 60 years.

> Silba used to have a population of about 1,100, and now there are 240 of us. They have all left—for Rijeka, Pula, Zagreb and Zadar. —male, aged 78, Silba, 2008

> When I was a child, there was a doctor and a dentist, and everything else in Nerezine. Before the Second World War, there were about 2,000 residents. Today there are 380, perhaps 400. And it's still okay. There used to be 300 or 400 in Sveti Jakov, and now there are less than 100. —female, aged 69, Lošinj, 2007

> There used to be 2,000, and today there are just 200 of us old folk. They all got flats in Šibenik and left. —male, aged 91, Zlarin, 2011

After the Second World War, the islands finally came under Croatia, one of six republics in the Federal People's Republic of Yugoslavia, later the Socialist Federal Republic of Yugoslavia. After the peace treaty between Yugoslavia and Italy was signed in 1947, mass emigration from Istria, Rijeka and the islands began, as well as from the parts of Dalmatia that had previously been part of Italy. The emigres were given a choice between two nationalities and chose Italian.

One woman recalled the following:

> There were only a few folk left, many had gone. I remember it like today. The 10th of July 1950, I was 12, and I cried all day long because all my friends from Sveti Jakov had left. There must have been six or seven families from Sveti Jakov who went with all their children. All my friends went to Italy. And stayed there to live. Some of them come now to visit in summer or their children come. Some sold their houses, some did not. And later, some went to America. —female, aged 69, Lošinj, 2007

The socialist order and planned economy were proclaimed. These new circumstances did not halt emigration from the islands, but the destinations changed. Now people moved mostly to coastal centres, which were being rapidly industrialised, or, to a lesser extent, to Zagreb and other parts of Croatia. During the first decades of the new state, it was not easy to leave the country legally, so emigration abroad became rarer. In this era, economic development on the coast concentrated on Pula, Rijeka, Zadar, Šibenik, Split and Dubrovnik, large coastal cities that became centres of development. They had industrial plants and traffic, and trade grew, and more housing was rapidly built, attracting people from surrounding areas and also from the islands who had always gravitated toward the cities. The islands remained on the spatial and economic periphery, with no significant financial investments or economic plans. This prompted further emigration.

Since their islands were poorly developed economically, large numbers of islanders found work in industrial centres in nearby mainland cities. At first, they commuted daily to work; however, many moved permanently to avoid the poor transport connections that made travelling difficult. The outer islands, from which it was almost impossible to commute daily, emptied quickest. During the most intense period of industrialisation of the mainland coast (1961–1981), they lost almost half of their inhabitants (15,595 in 1961, and 8,348 in 1981, according to census data). The islanders' decisions were made easier by the availability of social housing. The state directly influenced emigration from the islands by insisting on macroeconomic policies to stimulate and concentrate industrial production in the cities, while neglecting regional development, particularly on the islands. It ignored the need to invest in island housing and good traffic connections between the islands and mainland to facilitate daily commuting.

> During all the stages of socioeconomic development after the Second World War, the islands were on the margins of social interest, so it might be claimed that the depopulation process was prompted by the neglect of the former state regarding the valorisation of island resources and developmental potential, which led to further disproportionate conditions between life on the islands and life on the mainland. (Lajić 1997, 23)

In the 1960s and 1970s, periods of the greatest economic and demographic growth in the coastal cities were also periods of the greatest depopulation for the Croatian islands. As a result of emigration during a long period of time, mostly by young people of a fertile age, natural decrease occurred on the islands, further reducing the population.

Since the early 1970s, agriculture as the main economic activity on the islands has given way to more productive sectors. Industry arrived on the larger islands, followed by tourism, while the smaller islands exclusively took to tourism (Podgorelec and Klempić Bogadi 2013). Given the structure of the population, tourism began to develop when most young people had already left the islands. Tourism strengthened the island economies significantly and led to the construction of communal infrastructure, hotels, campsites and managed beaches on the larger islands; however, demographic effects were still lacking. Contrary to expectations, tourism did not encourage demographic stabilisation on most islands, as shown by demographic indicators: Natural change, which was 43 from 1962 to 1971, fell to –2,026 between 1972 and 1981; the migration balance from 1961 to 1971 was –12,209. It recovered somewhat between 1971 and 1981, at –10,769 (Lajić 1997).

Depopulation Beyond the Statistics: 1991–2011

The islands continued to experience population loss until 1991, when the census revealed growth that continues today, at least statistically. In the early 1990s, the political order changed, and Croatia emerged from the Yugoslav federation as an independent state. The ministry responsible for development and reconstruction established a Council for Demographic Development and a Council for Island Development. In 1997, the National Programme for Island Development was produced, and in 1999, the Islands Act was adopted. None of these actually led to significant changes in policies toward the islands, nor did they encourage investment in the economy (apart from investors in tourism, who needed no encouragement). No changes in policy resulted in no changes in the structure of the island economy, which continued to rely heavily on tourism. The quality of tourist accommodation and services varied widely from island to island. In general, economic and spatial development were haphazard, and in the last 25 years, significant stretches of the island coastlines have been ruined by the construction of holiday homes and apartments, often without the necessary building permits. "The state has only taken care of the islands so far in terms of maintaining regular ferry lines, which at least allow island life to continue at the weekends and in summer. This is not enough to revitalise the islands, which require specific island development policies" (Poljičak 1995, 82).

On the other hand, censuses reveal that there are more and more islanders. According to the last census in 2011, there were 47 inhabited islands with 124,955 inhabitants. Krk led with 19,383, and Male Srakane was in the rear with only 2 inhabitants (see Appendix A). Of the total number of islanders, 62.7% live on the near islands, 31.9% on the channel islands and 5.4% on the outer islands.

The reason for apparent demographic growth is also due to the inclusion in the census of the population that only lives on the islands for part of the year, mostly in the spring and summer months. These people mostly live elsewhere and own holiday homes on the islands but because of various benefits choose to register their official permanent residences there. This increases the overall population statistically. Research on the Šibenik islands has shown that as many as 30% of those permanently registered are in fact only occasional inhabitants (Podgorelec and Klempić Bogadi 2013).

The main reason for registering residence on an island is to qualify for residential tax and avoid the extra tax incurred by owning a holiday home. These fictional residents also have the right to free or subsidised ferry travel and other benefits. For instance, the 1981 census showed that there were 4,055 residents engaged in temporary work abroad, while in 1991, there were 10,342. Since there were no mass emigrations abroad during the 1980s, more "outsiders" were recorded by the census (Lajić and Mišetić 2013). In 1991, most of the fictional island population consisted of people working abroad and emigrants, while in the 2001 and 2011 censuses, holiday homeowners dominated (Lajić and Mišetić 2013).

Natural population trends on the islands raise doubts about the credibility of statistical census data, since any increase in the population through migration trends should at least counteract the low birth rate (Lajić and Mišetić 2013). Since natural trends remain negative, it may seem that demographic growth is the result of higher immigration and lower emigration, and that the islands have become attractive immigration destinations. This, however, is not the case. If it were,

> many island problems would not exist, and increasingly emphasised concerns about the islands, as well as the Islands Act, would not need to be expressed in order to achieve the basic goal of preventing further island depopulation or the actual extinction of the population. (Klempić and Lajić 2005, 341)

In spite of statistical indicators that speak of positive demographic trends, depopulation of the small islands continues, and they are still an area of emigration. In comparison with larger islands, they are in a significantly worse economic situation. There are too few permanent jobs and the tourist season is short, so that the working-age population does not have guaranteed employment, and this continues to prompt young people to move away. In addition, there are fewer and fewer educational institutions and no organised extracurricular activities for children. The main factor preventing young families from staying on the islands is the closure of schools, primarily on small and distant islands, while it is impossible for youngsters to commute daily to mainland schools due to inadequate transport connections. So, many parents move to the mainland with their children. This is often the precursor to permanent resettlement. The lack of kindergartens and schools on the islands also discourages families with school-age children from moving to the islands.

Because pupil numbers are so low, the island schools are constantly threatened with closure. An island school is important not only in terms of education, but also for the community itself. A school symbolises life and the survival of the community, as one of our respondents explained.

> I love Zlarin, but it needs 30 young families, people with jobs, a kindergarten, a school—and for there to be no difference between winter and summer. To live all the year round. —male, aged 47, Zlarin, 2011

Intense depopulation and ageing on small islands have led to the population dying out. On the other hand, in the two last intercensus periods, the population on bridged islands (with the exception of Murter) has grown considerably (26% from 1991 to 2011). Bridged islands (Murter, Krk, Pag, Vir and Čiovo) are of course in a separate category. By eliminating their physical separation from the mainland—an objective state of insularity equal to spatial, economic or social aspects—their islandness has been deeply affected. Any permanent traffic connection with the mainland affects the overall lives of individuals and the island community. In debates about deinsularisation and preserving islandness that have raged only recently in relation to the Croatian islands, bridges are both praised and derided. But it cannot be denied that bridged islands retain their young, working-age populations and attract nonislanders to move to them.

Geographic distance from the mainland has proved to be a significant factor in the demographic development of the Croatian islands. Islands closer to the mainland and bridged islands have more favourable demographic trends (Nejašmić 1991; Klempić and Lajić 2005) and growing populations, while demographic strength decreases with distance from the mainland. Between 1991 and 2011, the islands closest to the mainland recorded demographic growth of 9.4%, while the channel islands recorded a decrease of 8% and the outer islands an even sharper decrease of 28.1%.

In addition, research on the Zadar islands has shown that there is a significant link between attitudes and length of stay and distance from the mainland. Zadar is the mainland coastal centre to which all the islands in the Zadar archipelago gravitate. The closest island is Ugljan. When they decided to leave, as many as 62% of Ugljan emigrants thought it would be only for a short time (Babić, Lajić and Podgorelec 2004). Those from more distant islands with worse transport connections often decided to go for longer. Only 20% of those leaving the outer island of Dugi Otok and only 25% of those leaving the channel island of Iž intended to return quickly (Babić, Lajić and Podgorelec 2004).

MIGRATION TYPOLOGY

The history of relations between the islands and the mainland mostly boils down to the history of migration. Migration, which was often the response of growing dependence on the centre, often revealed the islanders' degree of insularisation. Migration was undoubtedly linked to the islanders and their way of life. To demonstrate its influence on the structure of contemporary Croatian island communities, we created a typology of islanders classified in seven categories in relation to the history of migration. The proposed categories overlap to some extent with the experiences of inhabitants of islands off the European and African coastlines or in the Mediterranean and Atlantic (the Balearic Islands, the Canaries, Malta and Cyprus), as described in Russell King's typology (2009). We have purposely ignored some aspects of categorisation in King's typology (for example, the financial status of migrants) and considered others: the specific geographical position of the Croatian islands in relation to the mainland, their size and inhabitation, various sociopolitical orders throughout history, traditional culture and the level of economic development.

1. The first group consists of *nonmigrant inhabitants*, people who were born and have lived all their lives on the islands. They do not consider themselves migrants or people with migration experience. This group includes men who spent a year or 18 months away on military service, which was compulsory until 2008. It also includes people who travelled away from their island as tourists or received treatment in mainland healthcare facilities. For example, in research conducted on the Šibenik islands, a quarter of the respondents, mostly older than 60 years of age, said they had no migration experience (Podgorelec and Klempić Bogadi 2013). Research on the Zadar islands confirmed that the commonest reason for staying among nonmigrants was family, whether this meant parents who they were reluctant to leave or starting their own families. Some answers given by the older respondents included that the people never wanted to leave their island, that it never used to be the fashion to go but to stay with their own folk, and that the people never saw anything better so they never wanted to leave (Podgorelec 2008). Other reasons given, although less important, were the security provided by island life, the way of life and having a roof over one's head and a job (Podgorelec 2010). In this respect, there is no significant difference between near and distant islands.

   Although they did not migrate, this group was still exposed to migration influences through the experiences of their relatives and friends who left, and through communicating with immigrants and tourists. The experience of a respondent from Cres is common.

   > I was born in 1921, and lived with my mother in Podol. In 1924, my father went to America. He lived in Boston, Philadelphia and New York. And he died there. —female, aged 86, Cres, 2007

   Relationships with migrants affected the life of the islanders directly or indirectly in many ways. Since the island communities were so small, people knew one another well and were constantly in touch. They depended on one another on a daily basis. In addition, the development of tourism exposed them to external influences that they did not all assess as positive.

> Later, all those foreigners came. You can imagine what it was like for us. The newcomers couldn't accept our customs. It was hard. The locals left, and we stayed. Our way of life changed. It was different. It wasn't like life in the village anymore. —female, aged 78, Cres, 2007

2. There are many *returnees* on the islands, people who were born on them but spent part of their lives away before going back. Research has shown that most people in this category are retired, that is, they only decide to return to their island after retiring (Babić, Lajić and Podgorelec 2004; Podgorelec and Klempić Bogadi 2013; Bara 2013). Their numbers have increased since the late 1990s, when those who left during the 1960s and 1970s (the most intense emigration period) began to reach retirement age. For example, most returnee respondents in the research conducted on Korčula emigrated in their early 20s, in the 1960s and 1980s, mostly abroad or to larger Croatian cities (Marinović Golubić 2017). Returnee islanders have usually inherited family properties on the islands where they were born or built new ones with the proceeds of their work, which has made returning easier. Many of them made preparations to return while working and living on the mainland (Podgorelec and Klempić Bogadi 2013).

   "Emotional links were never severed, and respondents thought that time spent away from the islands was just a stage in their lives prompted by the need to find jobs, education for themselves or their children" (Bara 2013, 213). Regarding returning in retirement, the respondents often mentioned an idealised image of life in the island community. One person who returned from Switzerland said,

> The island is my home. I was forced to leave because there was no work, but I always dreamed of returning. Returning to this beautiful but raw landscape with the closeness I felt toward the neighbours among whom I lived. —female, aged 74, Zlarin, 2011

During the years spent as émigrés, many returnee islanders visited their island regularly and were involved in island activities, and according to their experiences, they never ceased to be part of the community that made them feel they belonged and were accepted. Nonetheless, there were those who were disappointed in the changes in their communities.

> People are strange, today. They have changed. The young people don't greet you. Before people used to spend more time together and help one

> another. Life has become quite different. Today, about 200 old people live here, of whom 20 are newcomers. —male, aged 77, Zlarin, 2011

However, there were also economic reasons for returning. There was the opportunity to make extra income from agriculture or tourism, in addition to a steady pension, which meant having a better standard of living on the island. In comparison with foreign pensions, those earned in the former Yugoslavia and Croatia are generally lower, which is another reason for returning. For example, the average pension in Croatia in December of 2019 was HRK 2,724,78 (Statistical Information of CPII, 2020, Croatian Pension Insurance Institute), which is approximately $407 or £312. Returning to their islands was a migrant strategy to ensure a better standard of living in old age. It also meant improving the standard of living for younger family members.

> We lived in Šibenik. My son was a tenant. When we retired, we left him our apartment and we returned to the island. —female, aged 62, Zlarin, 2011

In this way, many émigré islanders resolved crucial existential issues by leaving their children the mainland homes in which they had lived (Klempić and Podgorelec 2002; Podgorelec and Klempić Bogadi 2013; Bara 2013).

3. Émigré islanders who no longer have permanent residence on the islands also participate in the island community and make regular visits. Depending on the location of their permanent residence, they visit on weekends and holidays, and during annual leave, and, in some cases, spend several months a year there. We have named these people *circulating émigrés* or "swallows," because they regularly circulate between the island and their places of permanent residence. Many of them identify with their native islands and feel part of the island community, particularly those who are there almost every weekend.

> When I visit my island, I live life to the full. I don't need to make arrangements to see people. I walk out of the house and someone greets me immediately. There is an intimacy, like living in a commune.—female, aged 61, Zlarin, 2011

Examples of emigres who regularly visit their island are evenly distributed throughout all the Croatian islands, regardless of their size, populations or distance from the mainland. Although they mostly see themselves as occasional residents, some regard themselves as permanent residents. Almost all have property on their native islands, whether inherited or built under their own supervision, and the fact that they can freely come and go while maintaining their property in a way obliges them to do so regularly. Some come because of their parents, helping them to maintain the household and agricultural holdings, and keeping the olive groves and vineyards in good shape. Some do not plan to return permanently but maintain active links with the island, its culture and way of life. They are active participants in island events and celebrations, work campaigns and cultural events.

4. *Daily and weekly commuters* in both directions are people who travel regularly on a daily or weekly basis from their island to the mainland for work in the coastal cities or those who travel to the islands daily for work. Daily commutes are only possible from the nearest islands with frequent ferry lines and short journey times. People only tend to travel weekly from more distant islands. The number of commuters varies from island to island, depending on size and how much the economy has developed. The more economically developed islands, with an urban centre and/or developed network of settlements, have a lower rate of daily commuting between the island and mainland. For example, the town of Mali Lošinj attracts most commuters from the island of Lošinj and the nearby island of Cres, connected to Lošinj by bridge (Stražičić 1981). The town of Mali Lošinj is the largest Croatian island town, with 8,116 inhabitants (2011). It is rare to find daily commuters to or from the small islands, however. They are mostly teachers, doctors, postal workers and so on. On the larger islands, this type of commuter is more frequent, particularly during the tourist season, when many seasonal jobs are available. Commuting from the islands to the mainland for work or education has steadily declined during the last 50 years. It has almost died out on the small islands, since the age profile of the islanders means that there are almost no people left who might need to go to work or school in mainland settlements (Podgorelec and Klempić Bogadi 2013; Lajić, Podgorelec and Babić 2001).
5. *Newcomers or immigrants* are those who have moved to an island from other islands or the mainland. Among them are *marriage immigrants*, *labour immigrants* and *descendant immigrants*. Most are women who married on the islands or moved to their husbands' native islands when they retired (Podgorelec and Klempić Bogadi 2013; Marinović Golubić 2017). Work migrants moved and stayed there for work. Their numbers are considerably higher on the larger islands, where the economy is better developed and jobs are more plentiful in comparison to the smaller islands. Sometimes, the descendants of islanders, their children or their grandchildren born abroad move back to the islands. Their decisions are usually based on experiences of spending time on the islands and participating in the island communities during regular visits or holidays from childhood onward (Podgorelec and Klempić Bogadi 2013).

Another specific aspect of immigrant groups on the Croatian islands relates to refugees; former exiles and refugees can be found in all three subgroups. Hundreds of thousands of people were exiled from the occupied parts of Croatia and Bosnia and Herzegovina during the Homeland War in the 1990s. Those from Croatia were largely accommodated in hotels on the Croatian coastline and islands. Those from Bosnia and Herzegovina were given temporary accommodation in Croatia before moving on to other countries. Some of them found work and started families there, so they stayed.

6. We also find *lifestyle migrants* on the islands. These are people who were born in other parts of Croatia or in other countries who have come to live on the islands for part or all of the year. Their motives are usually to do with the desire to live in a more pleasant, cleaner and safer environment where there is a different rhythm of life, activities and values than in their previous lives, with the aim of improving their quality of life. They often see the islands as mystical, romantic places, ecologically preserved, far from the madding crowd of overbuilt, alien cities, where a quiet lifestyle and close communities still dominate. They usually base their decisions on previous tourist experiences of staying on their chosen island, usually in the summer, for rest and relaxation, entertainment and socialising. The words of one such newcomer from inland Croatia describe both his expectations and the difficulties of adapting.

> I am not an islander, but when I first visited Zlarin 35 years ago, I decided to come and live here in my 50s. I built a house and came with my family. The natural world here is beautiful. I enjoy the sense of peace, which can be almost deafening in winter. It is wonderful, but it is not easy. It is hard to fit into a new, different environment. It is hard to make friends. The locals keep me at arm's length. —male, aged 60, Zlarin, 2011

Migrant lifestyles are embraced by the retired and those whose work means their place of residence is immaterial. In contrast to the situation on islands in other Mediterranean countries, for example, Spain and Greece (Lazaridis, Poyago-Thetoky and King 1999; King 2009; O'Reilly and Benson 2009), there are relatively few lifestyle migrants on the Croatian islands. There are not enough houses and settlements for them to live in, so they do not tend to congregate and create new communities within existing ones. For the same reason, there is little tension between newcomers and islanders sparked by different lifestyles, home and garden designs, customs, social values or habits. Research on the Šibenik islands has shown that most immigrants have another home in their native land, alongside their island home, and move between the two depending on the season of the year. In answer to the question about the benefits to the island of tourists or newcomers, most islanders agree that newcomers are good for the island. They emphasise that they bring long-lasting values to the community, primarily by enhancing the social capital, but also by investing in the island space (renovating houses, etc.) (Podgorelec and Klempić Bogadi 2013).

7. *Intra-island migrants* are people who have moved within their island from one settlement to another. On the small islands, where there are usually only one or two settlements, this usually means moving from the interior to the shore. Whether the island is large or small, the direction is almost always the same. The main reason is the opportunities afforded by tourism as the main economic activity. Until recently, this has been located exclusively on the island shorelines. Another reason is the better transport connection between settlements on the shore and the mainland or mainland centre to which the island gravitates.

TOWARD AN ELDERLY COMMUNITY

Educational opportunities, jobs, favourable housing loans and the widespread availability of social housing were the main factors influencing islanders to leave their islands in the past 60 years. In socialist Yugoslavia, flats were allocated on a permanent basis to people who did not own a flat or house. The most frequent destinations were coastal towns and cities. This had a long-term effect on island economics and demographics (Podgorelec and Klempić Bogadi 2013). One of the most significant demographic consequences was the ageing of the population, particularly on the small islands. There are more women than men in the older population (see Figure 4.1), and there are many elderly people living on their own. The ageing of the island populations has been amplified by the high numbers of retired returnees. Although they have contributed to the stabilisation of overall population numbers or have at least slowed down population decline, they have also increased the number of elderly people in the overall island population and changed the social dynamics and needs for particular kinds of care in the community (Podgorelec and Klempić Bogadi 2013).

The average age of the islanders has continued to rise since the mid-20th century. This was confirmed by the 1953 census, when 15.7% were older than 60 years of age. In spite of statistical indicators of demographic growth from the 1990s to the present, the island population continues to age. According to the 2011 census, 17.8% of the island populations were young (0–19),

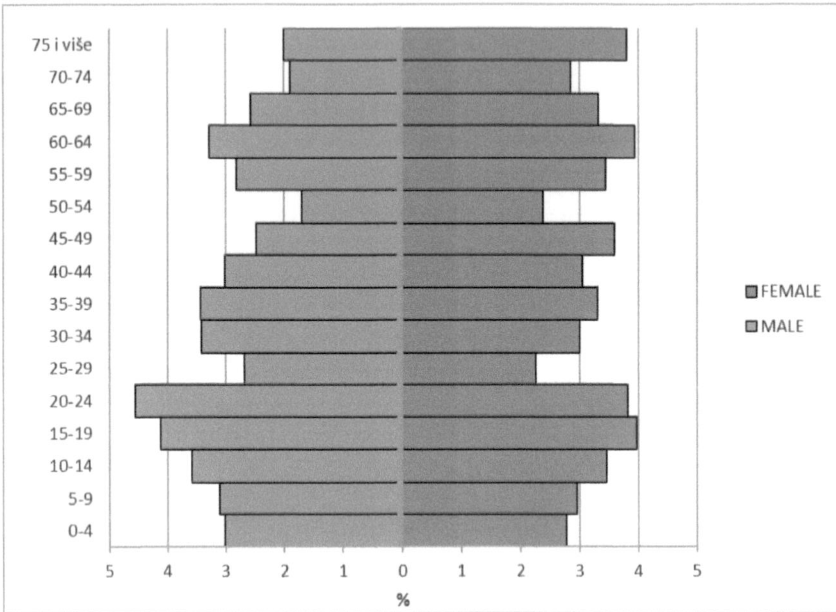

Figure 4.1a.

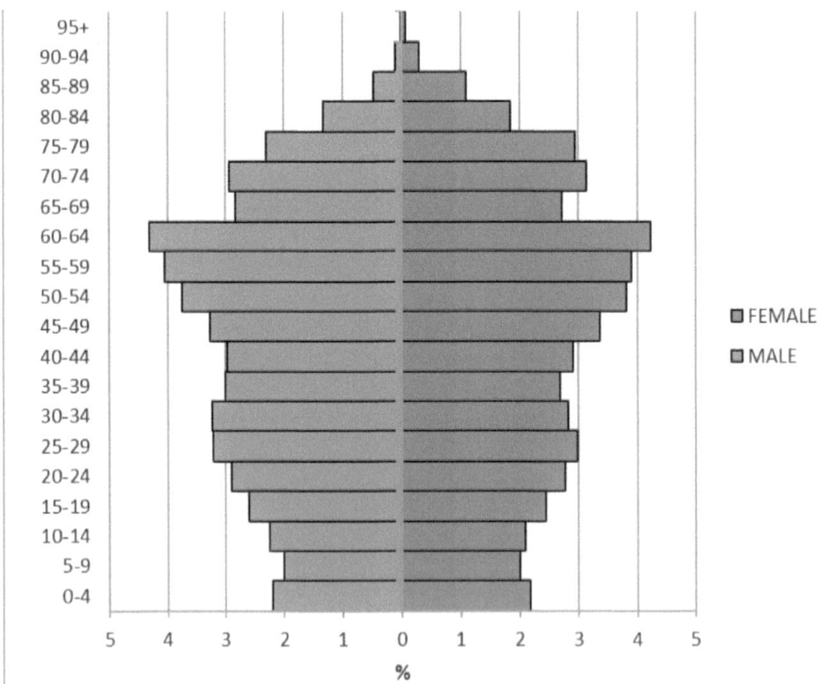

Figure 4.1b. Age–Sex Structure of the Population of the Croatian Islands in 1971 and 2011

and as much as 30.6% were older than 60 years of age. The average age was 45.1. There is no difference between the populations of islands close to the mainland and those farther away, so it is clear that proximity and better transport connections have not diminished demographic ageing processes (Nejašmić 2013). The situation on smaller islands is worse nevertheless; they have more residents older than the age of 80 (10%) than younger than the age of 19 (8.9%). The average age is 55.4, but half the population is age 60 or older, and this is reflected in every aspect of everyday life on the islands. The small islands have been the location of potential demographic extinction for decades, and some no longer have any permanent residents. Other settlements on more distant islands are also threatened, along with those in the interiors of large islands (Nejašmić 2013). One respondent on the island of Cres recalled the following:

> There was a shop and school in Ustrine. All the villagers had their own lambs and meat. Each house had its own. When I was born, there were 220 people here, but later, there were fewer and fewer year by year. Today, there are maybe 20. —female, aged 78, Cres, 2007

In the limited space of an island with an ageing population and few young people the "lack of certainty and desire to help each other means that social life is marked by solidarity and ultimate rationalisation" (Šimunović 1994, 240). Solidarity, the need to rely on one's own strength, rational organisation, carrying out daily chores, knowing how to spend one's free time and maintaining a high level of activity through old age are the basic characteristics of contemporary life in island communities (Podgorelec 2010).

Research has shown that although the cost of living on the islands is as much as 30% higher than on the mainland, the island environment enables the elderly to have additional sources of income (Lajić, Podgorelec, Babić and 2001; Babić, Lajić and Podgorelec 2004; Podgorelec 2008; Podgorelec and Klempić Bogadi 2013). Their meagre pensions form the basis, but if they are healthy and active, they can continue to be engaged in agriculture for their personal needs or bartering within the community. One respondent described a typical day:

> After lunch, you cast the nets and catch a kilogram or two, so you have something to eat. I am not interested in selling, but I do give fish to my friends. To the old people who have nobody, who can't go fishing. Mostly women. I give one a couple of fish to broil, another a couple to fry, and another a couple for a stew. And they give me cabbage (laughs). So we are even. —male, aged 74, Silba, 2001

Apart from working in the garden, fields and vineyards, or going fishing, some of them are also involved in the tourist trade. Most of the island populations think of gardening or olive-growing as hobbies or opportunities to socialise, spend time usefully and achieve material benefits. They often say things like this:

> I am pretty worn out now, I'm 80, but I can still do a lot. I still have the motivation, that's what's important. To feel useful. —female, aged 80, Cres, 2007

Apparently, growing old on the islands to a large extent means staying involved, active and productive.

The common opinion among islanders is that their island "has just become a big old folk's home where young people rarely visit" (male, aged 54, Dugi Otok, 2001). It is thus extremely significant that research into the quality of life on the Kvarner, Šibenik, Zadar and Split islands (Babić, Lajić and Podgorelec 2004; Podgorelec 2008; Podgorelec and Klempić Bogadi 2013) has confirmed that the population's active ageing dynamic structure (according to Boudiny 2013) occurs in three phases. The first phase, in the years leading up to retirement, is preparing to return to one's island by getting a future home ready and paying regular visits to establish oneself in the social network. The second phase consists of achieving an independent lifestyle, regardless of health, material or social problems, that will last well into old age. The third phase, when dependency becomes an issue, involves ensuring help is available from relatives and members of the local community (mostly neighbours).

An active lifestyle in old age is something to encourage on the islands whose populations are dwindling and getting older. Old people should be included on an equal footing in revitalisation projects affecting every segment of island life, from developing certain economic branches to improving culture and entertainment, with the basic aim of maintaining demographic strength. This is particularly important on small and medium-sized islands where old people are in the majority. As a group, elderly inhabitants have various interests, needs, functional abilities and health issues, and these must be addressed. Research into the quality of life among the elderly population of the Zadar islands confirmed that every fourth correspondent, regardless of their gender or marital status, the size of their settlement or the proximity to the mainland, was ready to be actively involved in projects to revitalise island life (Babić, Lajić and Podgorelec 2004; Podgorelec 2008). Hence, these approximately 9,500 islanders older than the age of 60 represent significant social capital.

In Croatia, care for the elderly in the family is above the European Union average. Approximately 17% of those aged 35 to 49 care for older relatives at least several times a week (European Commission 2015); however, the large number of old people and the small number of young ones in island communities means that special attention must be paid to organising formal types of care. On the islands, in terms of the population structure, there are simply not enough of these.

Islanders have access to specialist (consultant) and hospital health care in general and clinical hospitals on the mainland. There are care homes for the elderly and incapacitated only on the large islands (Lošinj, Vis, Brač, Cres and two each on Krk and Korčula), along with some smaller facilities (for 10 to 20 residents), for instance, the one on Dugi Otok. Altogether, there are beds for 564 clients, not all of whom are islanders. On the other hand, some islanders go to care homes on the mainland, but the greater problem is the lack of adequate noninstitutional care, the low level of civil society concern regarding this need on the islands and the insufficient numbers of volunteers to fill the gaps (usually relatives). Due to the emigration of people of working age, most islands lack professional caregivers for the elderly and incapacitated. There is a positive example on the Kvarner islands of Lošinj and Cres, where there is a high standard of accommodation for old people in two buildings. Yet, small settlements and small islands are completely lacking in noninstitutional setups that would provide help and care in people's own homes. Thus, everything depends on intergenerational support from relatives and good neighbourly relationships.

In spite of these shortcomings, in comparing their quality of life on the Zadar islands with that of their peers on the mainland (2001), almost half of the respondents (46.7%) said that elderly islanders lived better, a little more than a quarter (27.3%) thought standards were about the same ("Old age is hard everywhere") and about the same number (26%) thought life in other parts of Croatia was probably better (Babić, Lajić and Podgorelec 2004).

WHAT NEXT?

Although the Croatian islands have mostly been starting points for emigration during the past 100 or so years, they have also occasionally been immigration destinations. Respondents in various research projects (not just the sample on the Croatian islands) have listed the positive factors of modern island life that might attract newcomers, and they tend to be the same everywhere. For the small and medium-sized Croatian islands, they mention the quiet life, beauties of nature, clean environment, healthy lifestyle, security and close community as the greatest advantages (Babić, Lajić and Podgorelec 2004; Podgorelec and Klempić Bogadi 2013). By way of comparison, respondents on the Atlantic island of St Helena listed the same values but in a slightly different order: peace, close community, quality of the environment, freedom and security and way of life in general (Royle 1992).

A more peaceful life and less noise and stress are characteristics of life on the smaller Croatian islands, where there is usually no road traffic at all. The islanders are aware of the importance of an ecological approach to the island they live on and its natural features, where the sea is a crucial economic resource and an emotive, symbolic constant in their lives. On the other hand, the haphazard development of tourism and construction of holiday homes and apartments, often bypassing spatial plans and ignoring the lack of communal infrastructure, has devastated some coastlines and threatens to destroy the natural environment, which is a basic island resource.

It is important for islanders to feel they belong to a community, where almost everyone knows everyone else and mutual help is available. Old people whose children live on the mainland turn to relatives, friends and neighbours on the island when they need to. In spite of the changes to which modern island communities are exposed, there is still a high level of solidarity and closeness, which provides a sense of security, particularly for the elderly or those living on their own.

> It is good to know that your fellow-islanders will come to your aid. —female, aged 38, Kaprije, 2011

Islanders take it for granted that life on the islands is very safe. Research on the Šibenik islands (Podgorelec and Klempić Bogadi 2013) has shown that more than 90% of respondents felt safe living on their islands, mostly due to the low level of crime and the fact that in small communities, everyone knows everyone else. Insularisation, which limits the quality of life in some ways, also provides security.

> When the last boat leaves in the evening, we know that no more strangers will come. —male, aged 64, Zlarin, 2011

Alongside these advantages, there are still important economic and infrastructure problems (for example, the lack of jobs and education institutions) that prompt young people and those of working age to leave the islands. The physical barrier of the sea is a factor that limits access to all kinds of things that are readily available to people on the mainland: products, services, institutions, cultural and sporting events and so on. In the past, the islands were functionally sustainable, relying on local resources, but today, the inhabitants of small and medium-sized islands, in particular, tend to be oriented toward the mainland and depend on consumables from there. For this reason, one of the most important factors affecting the quality of the islanders' life is transport connections with the mainland. This is an important push factor affecting emigration—in the daily lives of the islanders, ferry timetables and frequency are rarely synchronised with their contemporary needs.

Since they are aware of the fragility and smallness of their communities, the islanders today are more prepared to accept changes introduced by foreigners or nonislanders. Above all, they are prepared to welcome anyone who wants to move permanently to the islands (along with occasional visitors and tourists), in whom they see hope for demographic and economic survival. At the same time, the indigenous islanders expect newcomers to accept the basic values of the island communities and not endanger the way of life they discover there.

REFERENCES

Babić, Dragutin, Ivan Lajić, and Sonja Podgorelec. 2004. *Otoci dviju generacija*. Zagreb: Institut za migracije i narodnosti.

Bara, Mario. 2013. "Povratne umirovljeničke migracije na hrvatskim otocima." *Migracijske i etničke teme* 29, no. 2: 201–24.

Boudiny, Kim. 2013. "'Active Ageing': From Empty Rhetoric to Effective Policy Tool." *Ageing and Society* 33, no. 6: 1,077–98. doi:10.1017/S0144686X1200030X.

Cukrov, Toni. 2009. *Mali otok pa se ljulja, I i II*. Šibenik: Znanje.

European Commission. 2015. "Country Report Croatia 2015." *European Commission*, https://ec.europa.eu/info/sites/info/files/file_import/cr2015_croatia_en_0.pdf (accessed 10 December 2018).

Friganović, Mladen Ante. 2001. "O demografskoj problematici hrvatskih otoka na Jadranu." *Sociologija sela* 39, no. 1/4: 37–59.

King, Russell. 2009. "Geography, Islands and Migration in an Era of Global Mobility." *Island Studies Journal* 4, no. 1: 53–84.

King, Russell. 1999. "Island and Migration." In *Insularity and Development: International Perspectives on Island*, edited by Emilio Biagini and Brian Hoyle, 93–115. London: Pinter.

Klempić, Sanja, and Ivan Lajić. 2005. "Prirodno i mehaničko kretanje stanovništva priobalnih, kanalskih i pučinskih otoka." In *Zbornik radova 3. hrvatskog geografskog kongresa*, edited by Aleksandar Toskić, 335–45. Zagreb: HGD.

Klempić, Sanja, and Sonja Podgorelec. 2002. "Migracije i starenje—pet iških priča." *Migracijske teme* 18, no. 4: 319–38.

Lajić, Ivan. 1997. "Suvremena demografska problematika jadranskih otoka." In *Nacionalni program razvitka otoka*, 13–32. Zagreb: Ministarstvo razvitka i obnove.

Lajić, Ivan. 1992. *Stanovništvo dalmatinskih otoka*. Zagreb: Consilium—IMIN.

Lajić, Ivan. 1989. "Utjecaj društveno-ekonomskih i demografskih promjena na iseljavanje s dalmatinskih otoka u 19. i prvoj polovini 20. stoljeća." *Migracijske teme* 5, no. 4: 307–24.

Lajić, Ivan, and Roko Mišetić. 2013. "Demografske promjene na hrvatskim otocima na početku 21. stoljeća." *Migracijske i etničke teme* 29, no. 2: 169–99. https://doi.org/10.11567/met.29.2.3.

Lajić, Ivan, Sonja Podgorelec, and Dragutin Babić. 2001. *Otoci—ostati ili otići? Studija o dnevnoj cirkulaciji sa* šibenskih *otoka*. Zagreb: Institut za migracije i narodnosti.
Lazaridis, Gabriella, Joanna Poyago-Thetoky, and Russell King. 1999. "Island as Heavens for Retirement Migration: Finding a Place in Sunny Corfu." In *Small Worlds, Global Lives: Island and Migration*, edited by Russell King and John Connell, 298–324. London: Pinter.
Marinović Golubić, Marica. 2017. "Doseljavanje na otok—suvremene migracije na otok Korčulu." *Migracijske i etničke teme* 33, no. 2: 115–41. https://doi.org /10.11567/met.33.2.1.
Nejašmić, Ivo. 2013. "Demografsko starenje na hrvatskim otocima." *Migracijske i etničke teme* 29, no. 2: 141–68. https://doi.org/10.11567/met.29.2.2.
Nejašmić, Ivo. 1991. "Depopulacija istočnojadranskih otoka." *Migracijske teme* 7, no. 1: 77–99.
O'Reilly, Karen, and Michaela Benson. 2009. "Lifestyle Migration: Escaping to the Good Life?" In *Lifestyle Migration: Expectations, Aspirations and Experiences*, edited by Karen O'Reilly and Michaela Benson, 1–14. Farnham: Ashgate.
Podgorelec, Sonja. 2010. "Migracija kao sudbina – hrvatski otoci i starenje stanovništva." In *Migracije i regionalni razvoj Hrvatske*, edited by Ivan Lajić, 141–62. Zagreb: Institut za migracije i narodnosti.
Podgorelec, Sonja. 2008. *Ostarjeti na otoku: kvaliteta života starijeg stanovništva hrvatskih otoka*. Zagreb: Institut za migracije i narodnosti.
Podgorelec, Sonja, and Sanja Klempić Bogadi. 2013. *Gradovi potopili* škoje*—promjene u malim otočnim zajednicama*. Zagreb: Institut za migracije i narodnosti.
Poljičak, Ivica. 1995. Šibenik na razmeđu: socijalnoekologijski ogled. Šibenik: Gradska knjižnica "Juraj Šišgorić."
Royle, Stephen A. 1992. "Attitudes and Aspirations on St Helena in the Face of Continued Economic Dependency." *Geographical Journal* 158, no. 1: 31–39.
Šimunović, Ivo. 1994. "Otoci u svjetlu socioekonomskih kretanja." *Društvena istraživanja* 3, no. 4–5: 451–65.
Statistical Information of CPII, Vol. 17, No. 12/2019, January 2020, Croatian Pension Insurance Institute (Statističke informacije HZMO, god. 17, br. 12/2019, siječanj 2020., Hrvatski zavod za mirovinsko osiguranje, veljača 2020).
Stražičić, Nikola. 1981. *Otok Cres: prilog poznavanju geografije naših otoka*. Otočki ljetopis Cres-Lošinj, 4. Mali Lošinj-Zagreb: Samoupravna interesna zajednica kulture općine Cres-Lošinj and Geografsko društvo Hrvatske.

SOURCES

Census of Population, Households and Dwellings 2001, Croatian Bureau of Statistics, Zagreb, CD ROM.
Census of Population, Households and Dwellings 2011, Croatian Bureau of Statistics, Zagreb, www.dzs.hr.
Settlements and Population of the Republic of Croatia, 1857–2001, Croatian Bureau of Statistics, Zagreb, CD ROM.

Chapter Five

The Identity of the Islanders

A Treasure Being Buried

Vladimir Skračić

I am an authentic islander from a time when most of the identifying characteristics of islanders were easily recognised. It has only taken one generation for most of these characteristics to be buried in forgetfulness; however, in spite of the tides of time, it is still possible to identify what is specific to islanders and islands from the relics of material and spiritual culture. The series of snapshots, scientific data and empirical observations that follow are the result of my experience and discoveries during linguistic research carried out on almost all the inhabited Croatian islands.

After the chapters giving an objective insight into near islands and the mainland, which always affects their existence to a greater or lesser extent, it is time to address the islanders themselves, who can often see the mainland from their windows. The archipelago is quite small, situated in a small Mediterranean Sea. The islanders have been largely unaware of the macrocosmos in which the microcosmos of their islands lies, surrounded by the sea, with the mainland on the horizon, and have built their way of life largely independently, learning to live with scant resources that have been difficult to sustain. Bordered by the island rim from which they could not travel far, they called their bays the "sea" and thought of islands further out as "distant." They also measured them according to their own standards: Islands that would be considered small elsewhere were, in their eyes, large.

They also think of the mainland as large. Since it developed faster, it increasingly lured islanders to leave their islands and head for easier, better lives. Many responded by moving to the mainland coast, although they often felt the pull back home from the first day. As a result, various push-and-pull factors work in both directions, and it seems the islanders cannot resist either. This dichotomy has characterised present and earlier islanders, and determined their identity to a great extent.

THE ARCHIPELAGO

The Croatian Adriatic coast and islands are exceptionally indented. The length of the island shorelines alone is more than 4,000 km. Indentation means there are a large number of islands (1,246), bays and promontories, and a diverse, once well-stocked underwater world. Situated close to one another, the islands act as a wall in front of the continental mainland, guarding it symbolically and physically from sudden change or threats from the open sea. On the other hand, their proximity to the continental shoreline has allowed a strong economic and cultural exchange to take place, so that in many regards, the identity of the islands is inseparable from the mainland. The territorial waters, with the epicontinental belt reaching to the sea border with Italy, cover almost the same surface area as the mainland territory of Croatia. In terms of understanding life on the Croatian islands, island practices and everything that has survived to the present day, it is important to indicate geographic factors first of all, which often vary widely from island to island yet are unique as a whole to the islands, since natural factors have affected the formation of the unique identities of each individual island and the entire archipelago.

Lie of the Islands

This factor is, of course, the most important. It is so characteristic that it has entered geographic terminology as the Dinaric direction, named after the mainland mountain chain that lies in the same direction. The Croatian islands are part of this chain so that the larger islands are elongated and parallel with the coastline. This has established the economic, cultural, ethnolinguistic and many other features of island microcommunities. A prominent example of this geographic disposition is the island of Pag, which has belonged historically to two powerful communes, in Zadar and Rab, and is divided between two counties, two dioceses and three linguistic subsystems. Its geographical determinant, with the available agrarian land, the direction of the main winds and disposition of its bays, has significantly affected the distribution and urban appearance of the settlements on the islands, particularly those on the island shores. Different rules have applied in the island interior.

Islands by the Mainland Coast

The islands and the mainland coastline form a unified coastal scene. The position of the islands by the coast and the proximity of the mainland have always been significant; many islands are so close that they have been joined to it by filling in the dividing channels or, more recently, by bridges. Are they still islands? Have they become what some call pseudo-islands? I will not attempt to answer this question here, but it makes sense to discuss to what extent new circumstances have affected the authentic identity of "real" islanders. One thing is certain: The new mainland population has exerted great pressure on the language and other aspects of the "old" islanders' lives and led to the accelerated loss of islandness that had been built for centuries.

Islands that fit this description today include Krk, Pag, Vir, Murter and Čiovo. This is not just a modern phenomenon; many towns and villages that are on the mainland today used to be on islands in the past. Rovinj, Nin, Tribunj, Primošten, Rogoznica, Trogir and Dubrovnik are just some of them. Perhaps being separated from the mainland, even by the shortest distance, was once considered a security factor making island life desirable. The contemporary appearance of coastal agglomerations confirms that, in normal conditions, choosing between the islands and the continent has almost always been to the detriment of the islands.

Large and Small Islands

Another factor that influences the characteristics of island populations is island size. The meaning of large and small, in global or Adriatic terms, has been explained in many places, including in the preceding chapters. For now, let us say that an Adriatic island is considered large if it has many settlements and most of the pertinent institutions (administrative, educational, medical, religious and so forth). The large islands provide their inhabitants with a good number of facilities, for example, security, education, economic activities and social and administrative services, while small inhabited islands function only as suburbs separated from them by the sea. This is particularly true of the Zadar and Šibenik archipelagos, both of which have many islands. The situation is similar on smaller archipelagos: Lošinj (5 islands), Trogir (2), Dubrovnik (3) and Vis (2).

Most of the villages on these islands belong to the nearest town in the administrative-political sense. They have the status of boroughs. This specificity is evident at the onomastic level, since none of the subarchipelagos, apart from the Elaphites (Dubrovnik subarchipelago), have retained their historical names. Instead, they are known by the names of the towns or cities on the mainland that lie opposite them. In the expert literature, they are called the Electrides, Mentors, Apsyrtides, Cretheus Islands, Liburnian islands and so on. The question is whether these are ancient names or those coined by geographers and historians.

The small Adriatic islands usually have only one settlement, sometimes two (Šipan, Koločep, Prvić, Iž and Rava). The island of Molat is the exception, with three settlements. The name of the island and the name of the largest settlement (or the one that was the most important at the time when the name arose) are usually the same. Sometimes the settlement gave its name to the island and sometimes the island to the first settlement. These rules applied to large and small inhabited islands; however, there are exceptions. Some islands have names that do not correspond to any of their settlements (Šipan, Koločep, Mljet, Brač, Šolta, Čiovo, Prvić and Dugi Otok). Why this should be is hard to say. In the case of Dugi Otok, for example, we know that it was called Pyzuch in the 10th century and that this name is retained in the Croatian version, not for a settlement, but for a plain and a bay (*Čuh* Polje and *Čuška*). A dozen settlements have been established since the Middle Ages, but none has taken the old name of the island; however, none of them were built on the site of ancient Pyzuch.

During periods when populations were growing and there was even overcrowding, the group of small inhabited islands comprised almost two-thirds of the inhabited islands (28 of 47). Today, barely 5,000 people are left on all the small islands together, and they are mostly elderly. This indicates that larger islands tend to retain some identifying characteristics more consistently. On the other hand, larger islands tend to relinquish their characteristics more quickly as they attract more interest from outsiders.

Extending to the Mainland and Neighbouring Islands

The karst soil composition and lack of surface water have made it difficult to guarantee the existential needs of the distinctly rural island populations throughout history. The toponymy thus often refers to things that are in short supply or different, rather than things that are plentiful and similar. Island toponyms often contain references to water, a crucial, scarce resource. The following are some hydronyms for land on islands that refer to water: Jadra, Olib, Rivanj, *Šipnata* and Vodenjak.

Islanders addressed the shortage of arable land by clearing spaces suitable for vineyards and olive groves. Good-quality soil and water for crops (cereals and vegetables) were only available in areas of sedimentary rock layers (flysch) and rare fields in the karst, so most cereals were obtained by bartering with mainland inhabitants. Short distances between the islands and the mainland facilitated this. The more prosperous island families bought estates on the continental coastline. People from the islands of Ugljan, Pašman, Murter, Prvić, Zlarin and Krapanj had property in the stretch between the coastal towns of Zadar and Šibenik. This kind of lifestyle, particularly in the waters around the small inhabited islands of the Lošinj, Zadar, Šibenik, Trogir, Vis and Dubrovnik islands, led to extremely active seafaring.

Most families living on large or small islands had their own boats, and the demand launched thriving boatyard businesses and related trades (blacksmiths, sail-makers, rope-makers and so on). Some small island boatyards survive today (Betina, Nerezine, Brbinj and others). Each fleet of wooden boats was inversely proportional to the island's size and arable surface area. On the other hand, and for the same reasons, there was more ground clearance on small islands than on large ones. The consequences are still retained today in the quality of the wines and olive products, and even more so in the miles of drystone walls marking the parcels of pastureland and forming barriers to prevent soil erosion on the terraced island slopes. Where drywalls can still be seen, and there are hardly any islands without them, they are unique reminders of the extreme scarcity of island resources and the attitude of the indigenous population toward them. A rule could be formulated here: The scarcer the island resources, the longer the drystone walls. On the island of Kornat, which has a surface area of 32.5 km^2, there are 323 km of drystone walls, while on Baljenac, near the island of Kaprije, the surface area is 0.14 km^2 and there are 23 km of drystone walls. The skill of building drystone walls was inscribed in the UNESCO Representative List of Intangible Cultural Heritage of Humanity in 2018.

ISLANDNESS: TOWARD A DEFINITION

These connections and disconnections between the mainland and the islands point to the complexities in identifying distinctive island identities. We should bear in mind a recurring question here: Are there real differences between the identities of islanders and mainland-dwellers, and if so, what are they? The fact remains that a large number of the islands lie close to the mainland coast, and their inhabitants originally came from the mainland. Existential resources were often the same on the islands and coastal rim. This is particularly true of the small inhabited islands. The agrarian cultures are more or less the same, and the fishing grounds, especially more distant locations, are equally accessible by boat. The religious world, which in the past was defined by one religion, Roman Catholicism, and local institutions did not differ much from those on the mainland and has not given rise to any significantly different lifestyles, traditional vocations or customs. Finally, most of the island population (except for far south of Dalmatia) spoke the same Chakavian dialect as the mainland-dwellers. The question must be asked as to whether there is any need to discuss the specifics of island identity, particularly nowadays, when many kinds of identity are being erased by media pressure and superimposed by anything and everything. I certainly believe the island world has remained unique in many ways and that its distinctive features have not yet vanished.

Living in isolation reveals a great deal about the character, mentality and identity of islanders. Although the word *isolation* is related etymologically to *island*, island isolation is much more complex than the geographical definition—a body of land surrounded by water. Do the people of Ugljan and neighbouring Iž share the same islandness? In the past, living in isolation was a given from which it was difficult to escape. Only a few islanders (seafarers, traders and boatbuilders) ever left for any reason. The advantages of the mainland coast were beyond the reach of most islanders. It is true that the people from islands close to the mainland and mainland towns were quite involved in urban life. The women often took low-paid jobs (laundresses, wet nurses and maids), while men went to markets to sell olive oil, wine and cheese; however, most islanders could not even achieve this level, due to the distance from their homes.

The strength of one's sense of belonging to an island would be tested when opportunities for a better life arose on the mainland. Migration to the mainland was implicit proof that for many, living on an island was not their own choice. It was just how things were. The fact that many who have left still swear to their islandness and remain enamoured of their islands (including me) does not change things. Even today, when communications with the islands have improved to an enviable degree and there are "organised emergency intervention units," and indeed the Islands Act, despite its inadequacies, guarantees a more secure life, islanders still moan about the lack of access to everything the mainland has to offer. The many conversations I have held while conducting field research have led me to conclude wryly that it would be best to move the mainland and its benefits to the islands. Desert island fantasies are all very well in literature but do not happen in real life. We love our islands, but we love the safety of the mainland even more. This schizophrenic attitude is part of our island identity.

Can islandness be measured? Is it possible to be an islander to a greater or lesser degree, and what are the measurement criteria? These are hard questions. In the case of the Adriatic, one thing is certain: The larger islands have retained their populations better than the smaller ones, and bridged islands have performed best in this respect, although they are not really islands anymore. On the other hand, elements of traditional island identity have been preserved best on islands with small populations. Here I do not mean islands that were once inhabited but are more or less abandoned today: Sveti Andrija, Male Srakane, Vele Srakane and Škarda.

It is clear that the degree of islandness cannot be attributed to the surface area, the number of inhabitants or proximity to the mainland. Many much more complex factors have created the island profile of today (colonisation, naval wars, raids, natural disasters, fishing innovations and seafaring are among them).

A hard life with few resources and great insecurity marked the history of the islanders (and for some, the present). Work and life were tough. They could only rely on what nature gave them, and the island coast and waters strictly defined this. Beyond these, they had little to hope for and were very aware of it. Thus, islanders have always been considered thrifty, tight-fisted even, and this is still true today. There are countless anecdotes and jokes about them. It is a sort of national pastime to remind the islanders of their thriftiness. It is not my place here to judge whether it is credible, but it is undoubtedly true that a constitutive part of the islanders' identity is a focus on themselves and their surroundings. Their attitude toward money is the consequence of their unique situation, with minimum resources and little or no chance of increasing them. So, it would perhaps be fairer to call them rational, aware of their limitations and that they could only use what they had on the islands or in the waters around them. That was not much, and certainly nothing could go to waste. Most jokes refer to the good old island days, but some are quite modern. One talks about an islander who could not go to sleep because he was worried that the light inside the fridge would stay on when the door was shut. Another went to fix the TV aerial on the roof, and when his wife asked if she should start making lunch, he replied, "Not yet. I might fall, and they'll give me lunch in hospital."

SETTLEMENT

The process of settling the islands was not linear and changed in intensity throughout different periods of history. The life of islanders after settling was not linear. The mobility they demonstrated by crossing the sea to reach the islands did not end after they settled. The first islanders landed and immediately proceeded to the interior to pursue a familiar activity—agriculture. Later, they moved to the shores and learned to fish, but they did not neglect their fields. Thus, they showed they were mobile and flexible in choosing occupations and gradually occupied the island space, engaging in activities that were viable there. Once they understood that the surrounding sea belonged to their islands, they were able to define their space and economies properly, and began to build insular communities and their identities.

For the first settlers, the Liburnians and their contemporaries, the islands were their appointed dwelling place. Their successors came as invaders, and the last, most numerous wave (the Croats) chose the islands as their permanent home. The population of the islands often fluctuated. Many were refugees or exiles from the mainland fleeing from occupying forces who had taken everything from them, killing many. The island historiography is full of information about the arrivals of vast numbers of exiles, sometimes entire villages. Escaping to the islands has happened throughout history, as can be seen from a specific example in the recent past. Twenty-nine years ago, when Serbian paramilitary forces and the Yugoslav Army occupied most of the Zadar hinterland and threatened the city, the number of pupils in the Zadar island schools quadrupled, and schools reopened on the most distant islands, where they had been closed for decades.

The first towns, in the full sense of the word, appeared on the Croatian islands in the era of Roman colonisation. Only Krk and Rab still have traces of their Roman names and remains of ancient urban structures. The island towns were small since the territory they administered was small. Apart from being important trading centres, they were also extremely vital for monitoring navigation in the Adriatic. Venice launched attacks on Zadar for centuries to gain control of Adriatic shipping. In the Fourth Crusade (1202), the Crusaders were actually encouraged to occupy and sack the city, although it was Christian and so forth. Surveillance of navigation was one of the colonisers' main strategic activities as well. Byzantium, for example, set up a system of surveillance and defence forts on the islands, many of which have been preserved—Svetac, Vrgada, Cres, Veliki Sikavac and Veliki Brijun, Rab, Gradina on Žirje, Tureta on Kornat, and so on.

Location

Settlements on the large Adriatic islands were arranged in two zones: the coastline, usually with one town (sometimes more, sometimes none), and the interior, with several small settlements (Blato on Korčula is one of several unusually large ones). The first settlements on the small islands were mostly formed in the Middle Ages and were farther from the coast. If the configuration allowed it, the island saddle was often selected. This was because the agricultural population did not particularly value proximity to the sea, at least not to begin with, while their fear of the open sea, where pirates and marauders were active, was disproportionately high. In any event, it was greater than the advantages of living on the island coast. Hill settlements had views of the sea in two directions, over two bays. If danger threatened one, they could flee to the other. There are many examples of this, including Premuda, Silba, Molat, Sestrunj, Vrgada, Murter, Žirje and Lastovo. Fear was not the only reason, however, and each case should be considered separately. It was important, for example, to ensure the safety of boats anchored in the bays, given the wind direction. Until the ferry dock was built in Krijal on the island of Premuda, the ferry used to dock there (SW shore) during the *bura* and in Loza (NE side) during the *juga*. Today, these settlements look quite different. What they have in common is growth toward the shore and on the shore.

Raiding the islands and carrying off the population on galleys were regular occurrences in the turbulent 16th and 18th centuries (the Cypriot, Candian, Cretan and Uskok wars). The churches in settlements on the coast were fortified (St Nicholas in Komiža, Holy Spirit on Šipan, St. Sylvester on Biševo, St. Mary in Vrboska and so forth), and defence forts were used as refuges by the populations in times of danger.

On the other hand, many smaller settlements in the interiors of larger islands gradually moved toward the coast. This trend began when the danger from the sea began to recede, and fishing, trade and possessions beyond their own islands began to play increasing roles in family businesses. In the past, the settlements in the interior had outposts on the coast, but today, with Croatia experiencing huge tourist and nautical booms, these harbours have become settlements in their own right. Their spatial features indicate how recent this phenomenon is: Communal needs have been overlooked, the streets are too narrow for the new volume of traffic, and there are no homesteads, cattle sheds, wells, threshing-floors, traditional village institutions like churches and schools, and such communal facilities as graveyards or public buildings.

Spatial Organisation

The cultural heritage on the Croatian islands, whether material or spiritual, is extremely rich and diverse. A literary tradition in Latin and Croatian flourished during the Renaissance on the large islands and in the towns. Even today, the islands of Korčula, Hvar, Pag, Rab, Krk and Cres boast architectural examples of exceptional beauty, jewels of the building styles of the day. Meanwhile, rural architecture in preserved villages, especially in the interiors, sets an example for modern builders, particularly regarding the rational use of space. The island mindset has always been obsessed with the need to exploit arable land, so settlements were always built on unfertile ground on a saddle or slope facing southwest toward the sun but hidden from the view of seafarers. The island builders' spirit promoted a modest, simple and functional approach to entire villages, houses, churches and graveyards. They used building material found on-site—stone for the houses and paving slabs, jetties and drywalls. The urban layout of settlements on the coast may have varied from island to island and village to village, and been quite different from the settlements in the interior, but it suited everyone because it met the spatial, economic and transport needs of all users. The same thing can probably be claimed for many settlements along the Adriatic mainland coast; however, nowhere is the harmony between the spatial organisation and user needs as evident as on the islands. An aptitude for building and spatial planning is one of the most recognisable characteristics of islandness.

Today, however, under the enormous pressure of construction for tourism development, the vistas of Adriatic coastlines and island settlements have become systematically depersonalised and uniform. Many no longer have the characteristics that made them recognisable for centuries. Weak or poorly implemented building regulations have resulted in such unsightliness in coastal urbanism that it is now impossible to reverse the negative consequences. Most of the Croatian islands and mainland coast are peppered with groups of holiday homes where no one lives except during the summer months. Several of the nearest islands have several times more houses than inhabitants (Čiovo, Murter, Vir and Pag). The situation has become absurd; the recognisable urban features of the island settlements, which have always been a tourist attraction, are being stamped out by unimaginative construction intended for tourists.

LIFE ON AND BY THE ISLAND SHORES

Sailing Through the Islands

For thousands of years, the coastline and islands have been part of a bustling maritime region. Trade routes from Greece and the Near East led along the Croatian coastline to the western Adriatic ports, primarily Aquileia and Venice. The Amber Road was probably the best known. The advantage of this route was that it offered good shelter from strong winds. Experienced captains could use the islands as havens in unpredictable weather. On the other hand, despite the many advantages of sailing along the island coastlines, there were also dangers in the fragmented archipelago: shallows, reefs, and tiny islands that took their toll. There are many toponyms commemorating accidents at sea. They are often named after the ships that sank or the travellers who perished (*Punta Puiza*, *Arbanež*, *Grego*, *Kandija*), and the Adriatic seabed is scattered with wrecks from many historical periods. Archaeologists have identified more than 1,200 so far. Thus, a unique, strangely symbolic museum of various cultures is to be found under the sea.

The sad company of the drowned has uniquely shaped the islanders. Boats, sailing and the dangers to which they were exposed affected the formation of individual and common features. Although they had no formal qualifications, the islanders were skilled at sailing their own boats in their own sea, which in their eyes equalled an ocean. Large vessels also sailed the Adriatic, often owned by islanders (from Silba, Lošinj, Hvar, Korčula and others). These ships were used for local cabotage (the *bracera* and *trabakul*) or transoceanic journeys (large sailing ships of all kinds). They were not specific to the islands or even the Adriatic but met Mediterranean demands for merchant vessels at the time.

Safety of Navigation

Safety of navigation was always a priority, as can be seen from reports on the Adriatic archipelago written a long time ago (Pseudo-Scymnus, Pseudo-Scylax, Strabo, Ptolemy, Pliny the Elder, the Anonymous Author of Ravenna, Porphyrogenitus and others). There are also wonderful linguistic proofs in the island toponymy. The need to communicate was just as great as it is today. Messages were sent from the highest points from which they would be most easily seen. These points on the islands were called *straža* (from the word for "watch"). Today most of the more significant islands in the Croatian archipelago have their own *straža*. There are more than 50 on the Zadar islands alone (Belveder, Čekanje, Javnjak, Ozrinj, Pudarica, Šentinela and so on)

Fire and smoke signals were used. Fire primarily indicated imminent danger but was also used for other reasons. The peaks above bays where shoals of large oily fish would often form were also patrolled according to an agreed schedule (Telašćica, Molašćica, Sakarun and so forth). Messages were sent by setting up agreed upon signals in visible places, and horns were blown to alert fishermen. This lasted until the 19th century, when Austria (of which Croatia was then an integral part) decided to improve security on the Adriatic sailing routes by building modern, imposing, manned lighthouses. There were about 80 large and many more small ones. Today, there are more than 700 lighthouses. Sailing the Adriatic at night is a real experience.

Family Boats

Small wooden family boats with sails were a particular part of the Adriatic islanders' identity. We tend to think they were all fishing boats, but this is not true. Most were small vessels, five to eight metres in length (the *kaic*, *gajeta* and *leut*), which could best be described as working boats. They were intended first and foremost for activities connected with work on the mainland or other islands, since almost every settlement had holdings elsewhere, and if not, boats were convenient for transporting heavy cargos from distant places on the same island (stone for construction, hay, timber for fuel, olives and grapes). Smaller loads were carried by donkey or the islanders themselves. In the case of islands with large holdings elsewhere (Murter, Hvar, Krk, Zaglav and Sali on Dugi Otok), the boat was the hub of the family business and often given priority treatment. The presence of boats influenced the urban vistas decisively (Hramina and Betina on Murter, Kaprije, Krapanj and Kukljica), as countless jetties were built, along with ramps and boatyards, frames for drying nets and other installations.

The boats themselves were robust, equipped with sails (usually lateen sails) and oars (usually four) to power them. They had few relatives of a similar constitution in the rest of the Mediterranean, and they can still be seen in places like Murter, Betina, Jezero, Prvić, Zlarin and Vrboska. They represent a unique feature of island identity. The islanders' loyalty to them has not died out, as is evident from successful attempts to revive sailing under the traditional lateen sail. One of these vessels, a *gajeta* from Betina called the *Cicibela*, is on the list of cultural goods of the Republic of Croatia. The very names of these boats tell a lot: *Breadwinner*, *Father Duje*, *Friend*, *Mother Anka* and so on.

Family Fishing

Family fishing on the islands formed part of the triad of survival (sheep, fish and olives, or earlier, grapevines). Islanders were family fishermen, so not many of them took part in commercial fishing (in the past, most fish caught for the market were small or large oily fish). Those that did usually joined grand forays when the fish were plentiful. While they were away on long fishing trips, they lived and functioned differently from the island community. Commercial (today we would say "professional") fishing boats were lighter and faster than working boats, and had more sails. There were defined rules on fishing, particularly in the more distant Adriatic fishing grounds (Palagruža, Sušac and Kornati), and the right to fish was usually allocated by ballot (*brušketa*). Small, family-based island fishing only took place around the coastline, near the settlements, and usually at night (so as not to waste daylight, which was for spending in the fields and pastures), using various items of tackle and methods. This type of fishing was emblematic of the island and coastal identity. It produced an extensive, diverse vocabulary of its own; however, there were linguistic problems involved in transferring the names used on the coast to inland fish-markets.

It is well known that words in this area are essential to identity, not only at the level of the archipelago, but also for individual communities. At the scientific level, these issues are resolved by taxonomic classification, but each community still calls each fish by its local name. An example is *Dicentrarcus labrax* (European seabass), which has at least five different names on the islands (*brancin*, *dut*, *agač*, *lubin* and *smudut*) and at least as many more variations on these names. It is said in jest that if you want to start an argument with an islander, you only have to insist that your name for a specific fish is the right one.

THE WEATHER

Forecasting

A particular characteristic of every island farmer-fisherman and the entire community is skill in forecasting the weather. Although it seems hard to believe today, in the age of smartphones and less direct involvement with nature, forecasting the weather for the island communities in the past was of crucial importance and formed part of specialised island knowledge. Doing it well meant they could organise their work on the island or at sea, sail and fish safely, and often their lives depended on it. They paid attention to everything when judging the next day's weather: the shape and movements of the clouds, the light of the stars and moon, the tides in the bays, the behaviour of animals in the pastures and the sounds and flight of the birds.

According to my own research and that of others, there are hundreds of words in the maritime vocabulary that refer to the strength and direction of the winds, sea currents, cloud types, shapes and sizes of bays and types of sea mist. This vocabulary has not found a place in the national meteorological lexis, although standard Croatian, not based on the Adriatic, has no adequate names for Adriatic winds. National forecasters only use two Adriatic names (the *bura*, from the NE, and the *jugo*, from the SE), but all other winds are named according to the compass. So, the *maestral* is known as a northwesterly, the *pulent/pulenat* as a westerly, the *lebić/garbin* as a southwesterly, the *oštar* as a southerly, the *levant-levanat* as an easterly and the *tramuntana* as a north wind.

Winds

An analysis of the directions and names of winds is perhaps the most effective way of showing the full complexity of island reality and the ability of the linguistic system to select the right names to accurately describe each group. Considerable variety arises in the area of slight, rare winds between the *jugo* (SE) and *maestral* (NW). There are many different names for the same wind, varying from island to island. Most are based on the place from which the wind comes: *oštar* from the south (Latin: *auster*); *japrk* from Africa; *lebić* from Libya and *hurijan*, *fore* and *garbin* from outside, from the southwest. But there are other names, too: *srdelar* from the word for sardine, a wind that is favourable for sardine fishing, and *smokvar*, from the word for fig, which facilitates fig-drying and so on.

If we know that the wind and its direction are always the results of the same atmospheric processes, then we might think that all winds always blow in the same places from the same direction and have the same names. But this would not be true. The island diversity and linguistic differences come to full expression here, because, for the islanders, the wind has never been a phenomenon to observe from behind a living-room window. It is a reality, and its variations affect life and behaviour. When islanders grow old and can no longer go to sea or out to the fields, they still listen to the weather forecast, although it will not affect their day much. It is an area in which they excel, no matter how old or disabled they may be, and a powerful testimony to how special they are.

LIFE IN THE ISLAND INTERIOR

Agriculture

Although surrounded by the treasures of the sea, the islanders have always been farmers first and foremost. In the first phase of island settlement, they were livestock breeders rather than labourers in the fields, but from the moment the settlements were established and island land redeemed (in the later 19th century), they turned to farming it. Vineyards held pride of place. That continued for hundreds of years until the early 20th century, when the vines were blighted. Before that, they produced high-quality, highly valued wines (some are still considered so today) that were served at the most prestigious European tables, according to archive reports. Island sheep's cheese has been famous since classical times. Other cultures (cereals, vegetables, potatoes, in particular, and fruit) varied from island to island, depending on the climate, availability of water, rural traditions and fields for planting. Yet, it was wine (and cheese and meat, to a lesser extent) that was the most recognisable island product.

When the vines failed, followed by a decline in livestock-breeding, their place was taken by olives, although not to the same extent on every island. In the age of intensive grape-growing, there were islands (sometimes even large ones, like Vis) that had no olive trees. The vine monoculture had dramatic consequences for the island demography. When they were blighted in the 1920s by Peronospora, and later even more by phylloxera, thousands of islanders left for the Americas. It was the first great emigration wave from the Adriatic. The same thing happened again, less intensely, in 1956, when there was a bitterly cold winter (in Adriatic terms) that withered the olives that had replaced the vines as another monoculture. Many people also fled the country at that time for political reasons.

Diet

On the whole, the island diet was straightforward, particularly that of the poorer islanders, who were in the majority. It was based on simple recipes using what the land and sea provided. Surplus fish were dried or salted for the winter, as were legumes (broad beans, peas and chickpeas). Salt was particularly important. The oldest way of collecting it was from salt flats after evaporation. In the later Middle Ages, every suitable island bay had a salt flat. Their remains can still be seen in the sea and in surviving toponyms: *Soline*, *Slanica*, *Solinski* (bays) and *Kučrine* and *Savrnari* (ruins of walls in the shallows).

Almost nothing was bought, but families would meet their needs by bartering their surpluses with mainlanders. Fresh fish were cooked in simple ways (stewed, braised or baked). Since there was a shortage of olive oil on many islands or the islanders sold most of what they produced rather than keeping it for themselves, many families would not have fried fish. Preparation would require as little time as possible, so that precious daylight was not wasted on trivial tasks; therefore, most labourers ate their main meal in the evening, particularly in winter when the days were short. Although they had limited ingredients, it would be impossible to mention all the many different ways in which food was prepared from island to island. Fish stews varied widely from not only island to island, but also village to village and even family to family. Each community had its culinary methods as a mark of its particular identity. Today, the dishes prepared from these ingredients are undergoing a culinary renaissance. Sadly, most of them are mere imitations, although the Mediterranean diet is on the UNESCO protected list. The national list of protected dishes includes *brački vitalac* (a lamb dish) from the island of Brač and cheese from the island of Pag.

INNER RESOURCES

Knowledge and Skills

Valuable examples of material heritage, rare items that have survived, are scattered among local collections and old taverns that have not yet been renovated for tourism purposes. Others live on in the memories of those who can still recall the time when the characteristic features of island life were present and these items were part of it; however, the knowledge and skills of the islanders are in great danger of being forgotten forever. There are island harbours that were once the feather in the cap of Croatian fishing fleets but where there are no traditional boats left today, let alone anyone capable of building them. On islands where flocks of more than 5,000 sheep used to graze, there are hardly any people left who know how to make cheese the old-fashioned way. In this respect, the lack of consistent policies is evident. The Ministry of Culture of the Republic of Croatia put traditional Adriatic fishing skills, beliefs and customs on the list of protected intangible heritage, but today most islanders are prohibited from fishing in the traditional way by legislation initiated by the Ministry of Agriculture.

Until the arrival of tourism on the mainland coast and islands, the island populations were mostly engaged in livestock-breeding, fishing and agriculture—often simultaneously. Fishing was organised professionally in several places (Palagruža, Kornati and Sušac, for example), as were seafaring (Korčula, Silba and Lošinj), pottery (Iž), coral-work (Zlarin), sponge-diving (Krapanj), salt harvesting (Pag), quarrying (Brač), building wooden vessels (Korčula and Betina) and timberwork (Olib and Silba). These trades and crafts provided decent incomes for those involved. There was no industry to speak of, apart from fish processing and stone-dressing. The lack of all kinds of things on islands that were then well populated forced the islanders to hone their knowledge and skills, often without any outside help. Many could turn their hands to many things. In addition to the tasks involved in the triad of sheep-farming, fishing and olive-growing (or grape-growing), they needed to know how to build, equip and maintain their boats, sew sails and learn to sail, weave and repair their nets, master fishing skills, make lobster baskets, preserve fish using different methods, make barrels, build drywalls around pastures, make cheese and oil and, last but not least, read the weather signs to ensure safe navigation. The list is almost endless, and it can surely be claimed that another component of island identity was the ability to tackle a large number of different skills. Perhaps it was the crucial component. I might call these resourceful people, at least for our purposes, *homo insularis*.

Religion

Homo insularis tackled spiritual questions as well. Religion played an indispensable, perhaps inevitable, part in the lives of the islanders. In their isolation, sailing the stormy seas and experiencing want and disease, there were few people or institutions to which they could turn. The Mother of God and patron saints were often resorts in times of trouble. The family and broader community remained tightly knit together. They supported one another in times of need, drawing inspiration from Christian values and island institutions (brotherhoods, mutual savings banks and guilds). Sunday lunch should be placed on the list of national intangible heritage for its unique cohesive power and the simplicity of the fare. Votive churches, altars and paintings in Adriatic towns; early Christian churches in the fields and chapels on the points of sailing routes; toponyms containing saints' names and such rituals as boat-burning and processions with statues of the Virgin are visible manifestations of their devotion. The calendar of economic and social events was formed around church holidays and festivals, and the food served often depended on these, too.

The oldest toponyms referring to saints' names are usually those of settlements and configurations on the mainland rather than the actual islands (Stomorica [St Mary], Sutvara [St Barbara], Stobreč [St Laurence], Supetar [St Peter], Sutulja [St Elijah], Sućidar [St Isidor], Supokrac [St Pancras] and so forth). Giving saints' names to islands is a more recent phenomenon, and most of these names have not undergone linguistic adaptations. The islands so named are usually uninhabited: Sveti Andrija (St Andrew), Sveti Ante (St Anthony), Sveti Grgur (St Gregory), Sveti Ivan (St John), Sveti Jerolim (St Jerome), Sveti Juraj (St George), Sveta Justina (St Justina), Sveta Katarina (St Catherine), Sveti Kliment (St Clement), Sveti Marko (St Mark), Sveti Nikola (St Nicholas) and so on.

Superstition

Religion was about faith, hope and the Christian code of conduct, but superstition was also widespread. "Do not go out at night," "Do not walk over a graveyard" and "Avoid encountering the bearers of bad luck" were pieces of advice taken seriously. Witches, werewolves, evil spirits, souls in purgatory, those who had returned to Earth repentant and all kinds of curses were part of the folk arsenal, and are still deeply rooted in island memories. On the famous fishing island of Iž it was forbidden to sweep the house; borrow anything; mention mice, rabbits or hares; ask a fisherman where he was going; wish a fisherman luck or spill olive oil on days when the fishermen set sail. These things heralded bad luck for the fishing trip. Of course, there was a system of "antidotes" to neutralise the effects. If the fishing trip went poorly despite all precautions and counteractions, the entire company would return to harbour, unload the tackle and equipment, take the equipment home, pull up the boat on coast, clean the boat thoroughly inside and out, relaunch the boat, load everything back onboard and venture forth again on another long trip often lasting many days. These procedures, it was thought, would get rid of the lingering effects of curses and ensure that the new venture would surely see better luck.

Folk superstitions are not specific to islands and indeed are found worldwide; however, it seems that hunters, whether on land or at sea, were particularly prone to them, so this included islanders by definition. There was a long list of objects and actions that would bring bad luck and others that would reverse them. At the same time, there were islands where certain priests overruled superstition altogether. I was born on one of them. The story goes that once, the people wanted the priest to bless their fields so that the mice would stop ruining their crops. The priest was heard to declare, "Why did you bring me? You need cats!"

Religious and Secular Heritage

The religious and secular heritage of the islands was first written down in Glagolitic, a unique Croatian script only used by the literate few known as the Glagolitic priests. Angular Glagolitic was typical of Croatia. Although it was not only used on the islands, and indeed was not used at all on some, it is often associated with the islands in the collective memory, particularly Krk and Cres, where some of the oldest written Croatian documents are kept (the 11th-century Krk Inscription, the 13th-century Baška Tablet on Krk and the 11th-century Valun Table on Cres). A distinctive expression of folk devotion was shown for Glagolitic singing, which was also not characteristic only of the islands. It was a specific type of choral singing in the vernacular that appeared in Benedictine monasteries in the 10th century and still survives in Dalmatia and on the islands. The Roman Catholic community in Croatia was the only one granted the privilege of singing in their own language before the Second Vatican Council. The islanders, whose religion permeates their lives, continue to nurture church customs today. The Hvar tradition known as "Following the Cross" (*Za križen*) is inscribed on the UNESCO list of intangible cultural heritage.

LANGUAGE

Island Dialects

It would be no exaggeration to say that language is still one of the most prominent indicators of island identity, although as people move away, local dialects are dying out or undergoing significant changes. Today, most island knowledge and skills are only preserved in words that are "not useful to anyone anymore," along with objects and actions that are no longer seen as having any purpose. Nonetheless, island dialects could be studied and at least partially preserved by adopting a rational approach. Those who engage in this, however, as those who engage in other segments of island culture, will need to bear in mind that they cannot analyse the linguistic situation on the islands separately from the linguistic situation on the mainland.

In the Croatian archipelago, there are still islands (for example, most of the Zadar and Šibenik islands) that differ markedly in the linguistic sense from the nearest mainland, regardless of inevitable diagenerational differentiation. There are some with very few linguistic differences from the mainland (for example, Pašman and Vir) and others that are particularly heterogeneous in the linguistic sense due to historical circumstances (for example, Pag). Finally, there are islands where contemporary linguistic processes have profoundly altered speech. On some, local dialects have suffered the strong influence of more prestigious variations spoken in coastal towns (for example, most of the settlements on Krk, Šolta, Brač and Korčula). On others, the Croatisation of predominantly Venetian areas spawned new dialects (the towns of Krk, Cres and Mali Lošinj). When studying the island linguistic identity, as far as it exists, we should always bear in mind that language is layered. It started long ago with languages that are not traceable today and continued with Dalmatian-Romance dialects spoken on the islands and coast from the early Middle Ages. They endured longer in some places than in others (on Krk, until the late 19th century). Croatian dialects overtook them but were themselves exposed to strong Venetian and Italian influences. These processes can be traced in individual lexical groups, from fishing and sailing to livestock-breeding and agriculture.

There is a constant principle in the linguistic evolutionary process: The more prestigious forms (literary, standard and media) always have a detrimental effect on the humbler ones (variants, dialects and local speech forms). Local speech forms tend to be restricted to specific semantic fields that are unknown or little known in the standard language. The national language of a country and its institutions usually have no linguistic policies for preserving local speech forms and integrating their lexis in the standard language. This is very noticeable if you want to look up the name of a fish, a wind, an item of fishing tackle or marine infrastructure in one of the leading Croatian dictionaries. Only standard words are offered, while entire swathes of characteristic island vocabularies are falling out of use and into oblivion. What is worse, there are often no standard words for particular objects or concepts, so surrogates are invented that no one uses in practice.

On the other hand, in the last 20 years, there has been rising concern among those who are conscious of the need to preserve the island linguistic heritage, at least in written (digital) form. Dozens of dictionaries of local speech, whether of islands or coastal settlements, and several monographs on island toponymy have been published, and a linguistic atlas that focuses mostly on exploring the maritime and fishing lexis is in preparation. Nonetheless, and in the face of the all-pervasive influence of education, radio, television and new media, island speech will naturally continue to survive, as it is the characteristic most deeply embedded in the islanders' lives. It can be heard even when they are trying to speak "correctly."

Toponyms, the Guardians of Island Memories

The oldest information we have about the islands and islanders comes from the names of the islands and configurations on them (toponyms). It is typical for toponyms to survive long after the reasons for introducing them have ceased to apply. For example, the toponym *Trata* is familiar to many islands. Its original meaning was a place where nets were hauled acoast (Latin: *trahere*—to pull). Although fishing using this method has not been practised for more than a century, the name and reminder of an activity that used to be vital to existence have survived. The same is true of *Japnjenica* (referring to limestone-burning), *Padrara* (a small quarry), *Osik* (a stone-walled cattle pen) and *Spansa/Steralo* (a place for drying nets, dating back to the time when nets were made of cotton and had to be dried after each use). Toponyms are always authentic witnesses to past times and sometimes speak more succinctly of the realities to which they refer than historical documents. On the other hand, many old toponyms that evoked the way of life on the islands and helped the people find their way around have been forgotten. The people who once used them are gone, while their descendants have given up living close to nature.

Language: A Witness to Island Economic and Social Relationships

Island toponyms refer to trades and professions that have died out but also relate to the distribution of economic activities and how it affected the population. The chief toponomastic division on the islands is between indigenous (Croatian) and imported (mostly Romance) expressions, of which Venetian was the most important, due to the Venetian administration of the Adriatic, which lasted several centuries. In spatial terms, the division often demarcates the interior from the coast. The further inland, the more toponomastic references bear Croatian names (*Vrh*, *Polje*, *Laz*, *Sad*, *Vlaka*, *Zgon*, *Liha*, *Volujak*, *Jama*, *Podanak* and so forth), while the configurations of installations on or near the coast often have Romance names. Some are even older (Dalmatian Romance or Greek). These include *Trata* (mentioned earlier), *Porat*, *Bisaga*, *Punta*, *Mul*, *Riva*, *Kažela*, *Magazin*, *Lanterna* and some that are harder to understand: *Vrgada*, Žut, *Murter*, Žirje, *Novalja*, Žman, *Savar*,

Sali, *Sestrunj*, *Maun*, and so on. The Croats migrated to the coast from areas that had had no contact with the sea. When they first arrived, they lived alongside the Romance peoples and later became subjects of Venice, adopting most of its maritime and fishing vocabularies and toponyms. The consequences are still evident today. Croatian island speech is packed with words of Romance origin, which have been fully integrated at most linguistic levels with the Slavic system.

On the other hand, toponyms relating to agriculture and livestock-breeding are mostly Croatian. This has been well documented and has led to another realisation based on the social structure of the lexis. Prestigious words for activities and those who engage in them (fishing, seafaring, and trade) were practiced by the wealthier classes, often foreigners. Humble words (agriculture, cattle-breeding) were practiced by the more impoverished, indigenous Croats, who were of course more numerous.

CONCLUSION

We are reaching the end of this (unfinished) list of islanders' identifying characteristics. I regret to say that their specific features, knowledge and skills proved worthless when the standard of living suddenly started to rise in the 1960s. Even though the quality of life improved, both on islands and the mainland, the islanders still deserted their homes in droves for the mainland. New factories, new hotels, new jobs and better educational facilities on the mainland sucked the island lifeblood dry. The mainland was simply perceived to be a better place with more potential for a good life, and perhaps it was, in objective terms; however, "the best" has always been the enemy of "the good"!

During that period, demographic regression affected all the islands except the bridged ones. The remaining populations grew older, and traditional labouring activities gradually died out, while at the same time there was a mass trend toward tourism and the service industry. The distinctive island identity grew fainter, and now, when it is too late to go back, we need to consider whether anything can be salvaged. In new circumstances (a new independent state, new communications and a better awareness of nature and tradition), we are beginning to understand that systematic neglect of the riches of the sea, its waters and islands may produce an unexpected benefit. Many islands are still rare oases of unpolluted, untouched nature, real treasure troves, and in a modern world burdened by pollution, they are assets of the highest value, right in the heart of Europe. Can their "backwardness" be seen as "progress"? Can modern explorers dig up the long-forgotten treasure and display it? If they succeed, will it prove to consist of more than the unpolluted sea and untouched nature? We have seen that there are invaluable, unused treasures waiting to be brought out: the language of a unique lexis, knowledge and skills that may no longer be needed but continue to fascinate and can still be learned, or at least written about, and a unique urban and spiritual landscape based on community and mutual respect. It hardly needs saying that the best places to discover these things are on the islands that have been developed the least.

PART II

Chapter Six

The Island and the City

The Zadar (Sub)Urban Archipelago

Anica Čuka and Josip Faričić

THE SUBARCHIPELAGO

Nissologists usually concentrate on exotic aspects when studying islands—their isolation, their remoteness or perhaps their environmental sensitivity (Ratter 2018). Isolation and remoteness are the causes of many island particularities, whether economic, demographic, ethnological, linguistic or cultural; however, the concept of remoteness from the nearest mainland or other islands is not perceived and interpreted equally in all parts of the world. The Croatian islands belong to the category of "near islands" and, in global proportions, are indeed close to the mainland; however, their poor traffic connections and position on the geographic margins of the Croatian territory and the functional margins of socioeconomic processes have led both islanders and mainlanders to view the islands on the edge of the Croatian archipelago as remote. Their lack of integration in the mainland infrastructure (for example, the lack of a water supply network) and frequent winter interruptions to traffic connections due to bad weather are factors that make the islanders feel even more remote from the mainland.

One of the factors that affects the perception of certain islands as "near" or "remote" is their functional link with the nearest urban centre on the mainland or another island. Speaking of remote rural areas, Royle says, "Being away from the capital and from other big cities can make places in truth or perception relatively backwards," referring to remote island areas as not only the rural periphery, but also isolated by their marine environment (Royle 2001, 46). It should be emphasised that nearness is a relative concept and is, of course, affected by the strength and complexity of individual islands' connections with the neighbouring mainland (see Chapter 1).

In global studies, the Croatian islands are not comparable with others that are much more distant from the mainland (Madeira, Iceland and similar islands) or small, remote oceanic island states. Because of their remoteness, such islands have developed their own economies, energy systems and other infrastructure, relying on their own resources. On the other hand, islands like the Croatian ones have relied to a great extent on the near mainland and become dependent upon it. To a much lesser extent, the mainland has depended on the islands, so that a special functional and quite asymmetrical relationship between the two has been formed.

The Zadar islands is the common name given to the subarchipelago facing the city of Zadar on the mainland coast (see Figure 6.1). They belong to the north Dalmatian island group (Faričić 2012), and 18 are inhabited. Their position, natural resources and specific socioeconomic development have influenced how individual island settlements and their transport and functional connections have been shaped. In the past, the islands were linked with one another by boat, but in the modern age, these connections have been neglected, and the islands are almost exclusively linked only with the mainland. The Croatian archipelago is widely dispersed, and the Zadar islands stand out in particular because of spatial fragmentation that hinders the synergy and complementarity of the inhabited islands. Although there are probably similar examples throughout the world, the functional and historical links and mutual influences of near islands and the neighbouring mainland have not been sufficiently covered in the available literature. The concept of an urban archipelago in the literature worldwide tends to mean a large city spread over several islands, like Copenhagen or Venice, with a network of narrow canals linked by bridges, creating a functional, unified urban structure.

In the past, the settlement of the Croatian islands depended entirely on natural resources. An island would be settled if it had fertile agricultural areas and water, quantities of which were barely sufficient in the karst areas. Later, when the islanders started fishing and turned to seafaring, it was important to find bays where they could build harbours sheltered from the prevailing winds. Along with agriculture and fishing, the backbones of economic development in the mid-20th century, the islanders mastered boat-building and made agricultural tools and fishing tackle, pottery, footwear, clothes and anything else they needed for everyday life that could not be accessed more easily on the mainland.

The Zadar islands did not have the prerequisites for developing urban settlements, in contrast to other subarchipelagos with large islands. They had meagre natural resources, small surface areas and thousands of years of functional connectivity with and dependence on their mainland urban centre, so only rural settlements developed on them. Pag was the only Zadar island where the settlements of Pag and Novalja emerged as urban centres; however, although closer to the mainland, Pag was still on the periphery, too far away to gravitate substantially toward the nearest mainland urban centres, Zadar and Biograd. Salt production, for centuries a historical and geographical activity based on a natural resource, along with intensive livestock-breeding, prompted the development of urban settlements on Pag. The island was also divided between the municipality of the nearby island of Rab, which covered the west part, and the city of Zadar, which covered the east part of the island. Pag was also divided between the Roman Catholic Rab Diocese and Zadar (Arch)diocese (Faričić 2003). The southeast end, which lies close to the mainland, was bridged in the late 1960s. The island of Vir, which is even closer to the mainland than Pag, was bridged in the 1970s. Although smaller than Pag, it quickly became a functional unit with the mainland and started losing its islandness (Faričić and Mirošević 2014).

Here, however, the focus is on the development of specific functional and transport connections between the city of Zadar and the islands that gravitate toward it. This research will be of use primarily to those dealing with near islands, which are quite different in terms of their socioeconomic development and islandness from other islands, due to the influence of neighbouring urban mainland centres. It is hardly possible to draw parallels with similar research elsewhere because, as far as we know from the available literature, no such research has been carried out. This book may serve as a case study in near island studies and an aid to studying similar phenomena throughout the world by analogy.

On the larger Croatian islands, small towns have mostly developed to which other smaller settlements on the same or neighbouring small islands gravitate (for example, Unije, Susak, Vele and Male Srakane and Ilovik gravitate toward Mali Lošinj). The islanders of Biševo and Sveti Andrija, the small islands most distant from the mainland, have gravitated toward the town of Komiža on the neighbouring island of Vis to form a single functional unit. The rest of the island space has been included in the economic and administrative urban functions centered in mainland towns. This has led to the formation of a specific spatial system—urban archipelagos—with two-way relations between the mainland and islands, particularly those linked to Zadar, Šibenik, Dubrovnik and Trogir.

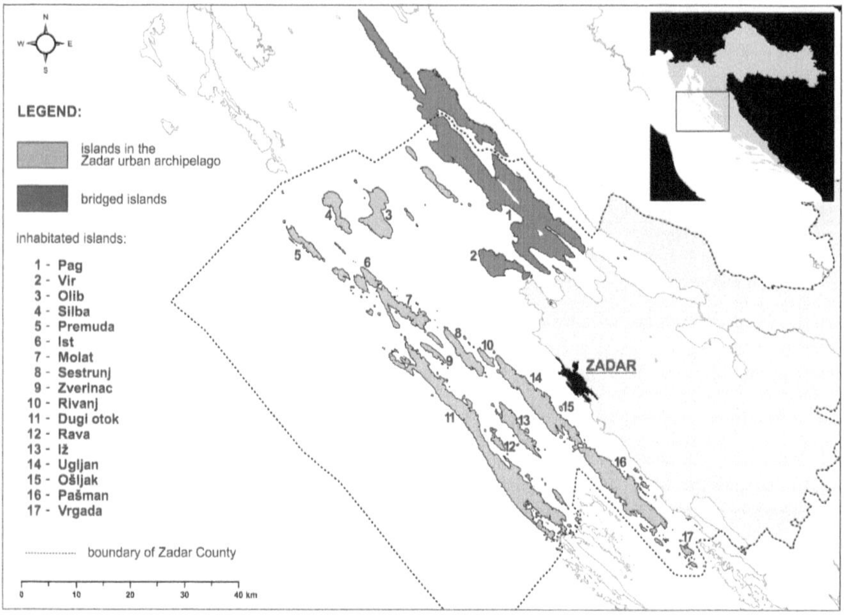

Figure 6.1. The Zadar islands

DEVELOPMENTAL FACTORS

The natural-geographic features of islands greatly influenced the socioeconomic development of the island space, particularly in the early stages of economic development. At that time, extensive farming was prevalent. Like most other inhabited Croatian islands, the Zadar islands were exceptionally poor in arable land, which hindered their development, reduced the extent of agriculture and limited the selection of plant cultures. Most large islands (Dugi Otok, Ugljan and Pašman) had fields with fertile soil, usually red soil. The arable areas on Ist, Premuda, Iž, Rava, Rivanj and Sestrunj were much smaller, which limited the production of cereals, vegetables and fruit, and also indirectly affected demographic development. The only cultures that produced sufficient quantities, or even surpluses, were olives and grapevines. In the past, drywalls dominated the island landscapes and were a particularly interesting ethnocultural phenomenon. The islanders built them to increase the proportion of arable land in the predominantly rolling, hilly relief. They cleared forested areas on the slopes and built terraces to prevent soil erosion. Thus, the landscape of terraced olive groves and vineyards was created. During the 20th century, agriculture was largely abandoned, and most of the cultural elements of the landscape once associated with arable areas are now lost under dense brushwood and maquis.

The Mediterranean climate gave way to an abundant growth of grapevines and olives, which became the predominant cultures on the Zadar islands. The islanders also had small vegetable gardens for their own consumption. Arable land was scarce, as was water, and these were the most constraining factors in terms of agricultural development, whether in the fields or with livestock. Most of the Zadar islands consist of limestone or dolomite. A great deal of surface water seeps underground through cracks in the karst rocks; therefore, no watercourses have formed, and there is no standing water. There are occasional island ponds, but these can hardly meet the need for water either in quality or quantity. Water scarcity is a prevailing challenge for the islanders that they have tried to resolve by digging wells (although the water is mostly brackish) or building rainwater collectors and tanks. Today, the lack of water still limits socioeconomic development on most of the Zadar islands. Water collection is low, particularly in the drier summer months when the demand is also higher, leaving the islands dependent on water ferried in from Zadar. Distance from the mainland, fragmentation and the lack of plentiful water sources that larger islands enjoy means that the Zadar subarchipelago suffers from a chronic water shortage. Only 4 of the 18 inhabited islands, Ugljan, Pašman, Vrgada and Vir, have been connected to the Zadar regional water supply network by underwater pipelines.

The sea acts as means of communication and a source of life, and is thus, understandably, the most vital natural resource for island development (Božanić 2020). The relationship between the islanders and the sea has for centuries been one of the most crucial components in valorising the sea and coastal areas. Fishing is the most obvious example. The first documented mention of fishing in the Zadar area dates back to the 10th century. Seafood consumption can be traced to even earlier; along with human remains, archaeologists have found the remains of marine organisms eaten in prehistoric times on the Zadar islands. When the islanders completely depended on natural resources, fishing developed well in settlements that had little fertile land or were close to rich fishing grounds. Kali (on the island of Ugljan) and Sali (on the island of Dugi Otok) have long been fishing centres in the Zadar archipelago. For many people in other island settlements, fishing was an important secondary, if not the main, economic activity. Almost all the other islanders fish as a hobby or for sport, which means they generally have fresh fish for consumption.

When agricultural development began to define the islanders' existence, most settlements developed in elevated positions, often next to the most fertile zones. Islanders rarely owned land, as it mostly belonged to wealthy aristocratic families from the mainland or the church. A dual system of land management developed on the Zadar islands: 1) the *colonat* system, which functioned for centuries, whereby the islanders were tenants working the land and delivering part of their crops and other products to the landlords, and 2) the system whereby many islanders owned land and worked it exclusively for their own needs. Apart from cereals and vegetables for personal consumption, they produced other things in larger quantities, mostly wine and olive oil, to sell on the markets. In the late 19th and early 20th centuries, the Zadar landowners allowed their island estates to be redeemed. They were mostly bought by the islanders who tilled them, and with that, the centuries-old *colonat* system came to an end. The islanders also bought land on the opposite mainland shore (for example, people from the island of Ugljan bought land in Zadar and the Zadar coastal area), which strengthened functional connections with Zadar. This was the era when the islands attained their maximum agrarian population and the island estates were no longer capable of producing enough food for the fast-growing population.

The 19th and 20th centuries included notable periods of crisis on the Zadar islands that slowed and, in some cases, halted their social and economic development. At the turn of the 20th century, they entered a demographic transition when the death rate slowed, while the birth rate remained high. The demographic decline only began, particularly on the smaller islands, after 1921. In 1857, there were 6,536 islanders on the small Zadar islands (excluding Pag, Dugi Otok, Pašman and Ugljan). The number rose to 9,426, the demographic maximum, by 1921. From then onward, it dropped continually. Two pests, Peronospora in the mid-19th century and phylloxera in the 20th century, ravaged most of the vineyards and brought calamity to families who had lived from selling wine (Čuka, Mirošević, Faričić and Graovac Matassi 2017). The islands experienced an economic slump that prompted the first intense wave of emigration in their history. During the Second World War, many refugees made the islands their homes and temporarily boosted the population, including the Zadar islands, which in turn had their highest populations in the postwar years up to 1948. When the refugees returned home and the islanders started to emigrate, the population plummeted and produced a negative birth rate (Graovac 2004).

Given the specific natural-geographic conditions and dependence on natural resources, special links have been built throughout history between the Zadar islands and the mainland. Since they could not develop reliable urban centres, the islands depended on relations with the mainland. This connectivity shaped a particular system—the urban archipelago, with the city of Zadar as its centre. The system is perhaps best described as a core-periphery model—the core is Zadar and the periphery the rural island area. The periphery can be gradated, and as a rule, the degree of peripherality (ending in almost complete marginalisation) increases with the distance from Zadar.

THE FUNCTIONAL CONNECTIVITY OF THE ISLANDS AND MAINLAND

Sixty-eight settlements have developed on the 18 inhabited Zadar islands (Faričić 2012). For centuries, the only urban one was the town of Pag on the island of Pag. In more recent times, thanks to tourism, the village of Novalja on Pag has developed into another island urban centre. According to the 2011 census, only six island settlements had populations of more than 1,000 (Novalja and the town of Pag on Pag, Vir, and Kali, Preko and the town of Ugljan on Ugljan).

Vir and Pag are bridged islands, and this has affected their functional integration with the mainland. As a result, their modern socioeconomic development differs greatly from the other islands (Faričić and Mirošević 2014). By correlating the distance, surface areas and populations of the unbridged islands from the mainland, it can be concluded that those farther from the mainland (particularly the smaller ones) have experienced less favourable economic and demographic development. Comparatively, those that are larger and closer to the mainland have succeeded in developing more settlements and increasing their populations at a faster rate. There has been an apparent link between demographic development and available natural resources. Islanders who did not have enough arable land on their islands had to acquire it on neighbouring islands or even on the mainland. Their settlements grew faster than those with small agricultural holdings. The people of Ugljan worked the land on the mainland around Zadar, while the people of Dugi Otok living in Sali and Zaglav possessed land on the Kornati islands and the neighbouring islands of Lavdara and Krknata, where olives are still predominant today. Agriculture claimed virtually every scrap of arable land on neighbouring islands and islets, and it was not unusual to find livestock, primarily sheep, left to graze on the smaller uninhabited islands.

Throughout history, the rural island settlements have supplied not only the islands, but also the city of Zadar, which has relied on certain island products. When the Zadar aristocracy owned the islands, annual accounts were kept to determine how much products the islanders were obliged to provide as their contribution to building up public goods and maintaining public areas in Zadar, as well as maintaining the saltworks on the island of Pag. Historical sources point to the exploitation of the forests for the functionality and maintenance of public buildings in Zadar. In the 18th century, the inhabitants of the larger Zadar islands were required to supply a certain quantity of wood and coal for fuel, along with stone slabs, lime, stone, sand and red earth, for the Zadar hospitals and other public buildings. Since not all the settlements could supply these items, lists were drawn up according to the capacity of particular islands. These lists indicate the existence and lack of different natural resources on the Zadar islands in recent centuries.

Guiseppe Lantana, a Zadar landowner, kept detailed diaries that included lists of duties the islanders had to complete during public works carried out in the 18th century. In the villages on the island of Pašman, islanders produced enough lime to become the sole supplier required to send lime to Zadar (Zadar State Archives 1759a). Other islands also produced lime but almost exclusively for local use. The larger, wealthier island settlements often had to provide various raw materials (Zadar State Archives 1759b). The Dugi Otok islanders in the villages of Savar, Božava and Veli Rat were required to provide wood for fuel for the public hospital in Zadar. The people of Sali, Žman, Luka, Brbinj and Dragove had to provide stone slabs. In 1751, seven island settlements (six on Dugi Otok and one on Rava) had to provide wood for fuel for the public hospitals (Zadar State Archives 1759c).

Archive records reveal that the Zadar islands also supplied the town of Pag, as well as Zadar. Documents dated 1744 and 1745 mention the number of logs the islanders were required to deliver for the reconstruction of the Pag saltworks. In 1744, this was 20,000, and altogether 11 island settlements were asked to cooperate in delivering them. In 1745, it was 31,167, of which 8,007 were large and 23,160 were small, so the settlements, according to their natural resources, were divided into two groups. The settlements on the islands of Molat and Ist had to prepare both sizes of log, while Rava and Božava on Dugi Otok only contributed a certain number of small logs.

In the early 19th century, the Austrian administration launched the first methodological, unified, detailed cadastral survey of the Dalmatian area. It included the Zadar islands and neighbouring mainland. In addition to the cadastre, the survey resulted in an exhaustive economic description of individual cadastral municipalities, which bore witness to the close links between the island settlements and Zadar. These documents frequently mentioned that the islanders placed their surpluses (mostly wine and olive oil) on the Zadar market (Čuka 2008; Faričić, Čuka and Colić 2010; Čuka 2013). In addition, Zadar attracted a specific labour force from its islands. The islanders found jobs in the city as auxiliary workers in households, for example, as laundresses, nannies and so forth. A solution to the problem of illegitimate, abandoned children revealed an intriguing link with the islands. There were several places in Zadar where women would leave such children. Foundlings were given for adoption, often to island families, who received monetary compensation for fostering them.

The links between the city and the neighbouring islands were therefore multifaceted and formed a two-way, spatial-economic system between the mainland and islands in the urban archipelago. The island settlements based their development partly on the use of island and maritime resources, mostly in the areas of agriculture, seafaring and fishing. Concurrently, the basic needs of the islanders in terms of services and administrative functions were met by the towns, primarily Zadar. At the same time, urban mainland markets received surplus island produce and other goods. Urban landowners gained material benefits from the islands. Towns used the islands as a pseudo-demographic receptacle for the labour force and biologically viable population to contribute to the birth rate. The islands were places of rest and recreation for the landowners, as is evident from the remains of country-dwelling architecture, which has maintained prominence in the landscapes of many Zadar island settlements.

Landowners and city institutions, and, in particular, the Archdiocese of Zadar, helped facilitate island–city integration. In each island settlement there was a parish; the exception was the island of Rivanj, where the parish had its centre on neighbouring Sestrunj. Regular pastoral and religious care throughout the centuries meant that there was close communication between the archbishop and his institutions, parish priests and congregations. In the past, the archbishop was a more frequent visitor than representatives of the city and county authorities—a tradition that still exists today on some islands. There were obligatory pastoral visits for special occasions, the administration of the sacraments, confirmations and various cultural events.

A PLACE TO WORK AND LIVE, OR FOR REST AND RECREATION?

During the 20th century, the Zadar islands experienced significant socio-economic changes. As a result of the decline in agriculture and fishing, and the growth of industry on the neighbouring mainland, along with changes in family planning, there were fewer and fewer islanders, and they were increasingly elderly. This had repercussions on the functional links between the islands and the mainland. The islands and neighbouring rural mainland hinterland proved to be an exhaustible demographic base, from which the population, mostly young and middle-aged people, relocated to mainland urban centres. In the 1960s, the city of Zadar experienced an industrial boom and attracted a significant number of islanders who never returned to their islands. From the late 19th century, the islanders went overseas, and in the second half of the 20th century, they also emigrated to European countries.

The Zadar area was also a focus of development in the service industries, which attracted both islanders and people in the hinterland. Agriculture declined and fertile regions were abandoned, so the agricultural island landscape soon began to change. The traditional cultural landscape began to disappear beneath indigenous Mediterranean vegetation, and an ageing, decimated island population could no longer produce enough food for local consumption, let alone for the Zadar market or elsewhere. The islanders' supply of food and other necessities began to depend on the neighbouring mainland. While two-way links had been a feature of the past, now the link was one-way as far as most islands were concerned, leading to dependency on the mainland. The exceptions were the largest and nearest islands of Ugljan and Pašman, and, to some extent, the outer island of Dugi Otok. Larger companies and public institutions (island schools, the shipyard in Vela Lamjana on Ugljan and the psychiatric hospital on Ugljan), and of course the daily circulation of workers, school pupils and students, maintained their relationship with Zadar.

Among other economic activities, the islands began developing tourism; however, it was seasonal and carried out by people who were originally from the islands but now living on the mainland. This can be seen from the disproportional growth in the number of apartments built for rest, recreation and other tourist accommodation in relation to the population level, particularly on the small islands (Faričić, Čuka and Colić 2010).

Traffic communication between the islands and the neighbouring mainland, however, has changed. While the islands were well linked in the past, currently, transport is almost exclusively organised between the islands and the main urban centre of Zadar. The islanders of Pašman and Vrgada are also linked to the mainland municipal centres of Biograd na Moru and Pakoštane. The reason for this is that the structure of the island economy has changed and islands do not produce goods to supply the other islands. For example, pottery was once produced on Iž and bought on other islands, and some settlements had boatbuilders (*kalafati*) who repaired the boats that were essential to island life (the last boatbuilders were located in two shipyards in Brbinj on Dugi Otok). The people of Pašman sold lime to the other islands; Olib was famous for its timber for boats, tools, furniture and so on. Today, these products are only available in Zadar.

On the other hand, smaller island settlements are demographically weak and on the verge of extinction. This is true of settlements on large and small islands, for example, Dragove, Savar, Soline and others on Dugi Otok (Graovac 2004; Čuka 2006; Faričić 2012). The population is predominantly elderly and not viable biologically. According to official statistics, the population is also generally economically passive. This has led some parts of the island space being transformed from a thousand-year-long tradition of continuous settlement to a place for rest and recreation. Nonetheless, the extent of the economic crisis and accompanying transformation of the landscape do not fully correspond to statistical categories. Although they are getting old and

Table 6.1. Population Trends on the Zadar Islands, 1948–2011

						1991				Index
Island	*1948*	*1953*	*1961*	*1971*	*1981*	*Total*	*Nationwide*	*2001*	*2011*	*2011/1948 1948. = 100*
Babac*	23	35	34	34	—	10	—	—	—	—
Dugi Otok	4,670	4,579	4,093	3,919	2,250	2,873	1,794	1,772	1,641	35.14
Ist	466	458	435	386	295	233	188	202	178	38.20
Iž	2,182	2,034	1,638	1,301	768	657	542	557	559	25.62
Molat	977	930	771	547	301	222	170	207	192	19.65
Olib	914	805	585	569	226	714	168	147	132	14.40
Ošljak	139	125	93	64	27	65	16	18	27	19.42
Pag	9,188	9,160	8,568	7,896	7,504	7,969	7,299	8,398	9,201	100.14
Pašman	4,310	4,223	3,761	3,730	3,324	3,349	2,622	2,711	2,844	65.99
Premuda	320	273	213	152	98	73	72	58	62	19.38
Rava	411	371	305	234	147	120	83	98	108	26.28
Rivanj	99	103	79	55	30	20	20	22	30	30.30
Sestrunj	382	373	322	354	96	123	59	48	45	17.78
Silba	514	444	397	339	198	221	206	265	286	55.64
Škarda	86	86	60	26	4	4	—	0	0	0,00
Ugljan	10,413	10,346	10,159	9,817	6,774	7,518	5,690	6,164	6,126	58.83
Vir	1,072	1,120	1, 069	959	866	860	777	1,608	3,032	282.84
Vrgada	538	513	449	408	311	236	223	242	240	44.61
Zverinac	158	161	153	146	96	59	53	48	42	26.58
Total	36,862	36,139	33,184	30,936	23,315	25,326	19,982	22,565	24,745	67.13

*In the last two censuses, the island of Babac was not recorded separately but added to the mainland settlement of Turanj in the Municipality of Sv. Filip i Jakov.

Sources: Korenčić 1974; Population Census of Households and Apartments 1991; Population at Home and Abroad by Settlements, Croatian Bureau of Statistics, Zagreb; Population Census of Households and Apartments 2001, CBS, Zagreb, Settlements and Population of the Republic of Croatia—Retrospect 1857–2001, CBS, Zagreb, 2005; Population Census of Households and Apartments 2001—Initial Results, CBS, Zagreb, 2011.

are officially inactive, many islanders still tend gardens, olive groves and vineyards that are close by and accessible by road. In addition, second-home owners, people who live in Zadar and the neighbouring mainland settlements but are originally from the islands, also look after their parcels and have built properties on them.

Table 6.2. Active and Inactive Population on the Unbridged Zadar Islands in 1971 and 2011

	Total Population		*Active Population (%)**		*Inactive Population (%)*		*Persons with Personal Income/ Pensions***	
Island	*1971*	*2011*	*1971*	*2011*	*1971*	*2011*	*1971*	*2011*
Dugi Otok	3,919	1,513	39.3	34.0	60.7	66.0	9.6	42,1
Ist	412	181	29.9	21.5	70.1	78.5	14.8	53.6
Iž	1,301	580	24.6	26.2	75.4	73.8	27.8	53.4
Molat	547	196	29.8	14.8	70.2	83.7	22.3	64.3
Olib	569	138	42.2	20.3	57.8	77.5	11.1	44.9
Ošljak	0	27	0.0	22.2	0.0	70.4	0.0	55.6
Pašman	3,730	2,483	41.8	36.6	58.2	63.4	6.2	32.6
Premuda	152	61	23.0	23.0	77.0	77.0	27.6	63.9
Rava	234	114	28.2	16.7	71.8	80.7	22.2	57.9
Rivanj	55	31	23.6	12.9	76.4	83.9	30.9	64.5
Sestrunj	354	46	27.7	23.9	72.3	73.9	10.2	69.6
Silba	339	273	22.7	34.1	77.3	65.6	35.1	47.6
Ugljan	9,881	5,411	32.0	38.2	68.0	54.2	17.0	33.8
Vrgada	408	225	31.6	20.0	68.4	80.0	16.9	43.1
Zverinac	146	43	47.3	14.0	52.7	86.0	12.3	62.8
Total	22,047	11,322	34.4	34.7	65.6	61.5	14.7	38.0

*Due to the proportion of active and inactive inhabitants in 2011, this does not add up to 100% on some islands, as there is also the category of "unknown."

**In the 1971 census year, pensioners were counted in the category of "persons with private income," while in 2011, they were counted separately.

Sources: Population Census of Households and Apartments 1971, Gender and Age – Part One; results by settlements and municipalities, volume VIII, State Bureau of Statistics, Belgrade, 1973: Population Census 1991; Population by Gender and Age, Croatian Bureau of Statistics, doc. 882, Zagreb, January 1994; Population Census, Households and Apartments 2011, population over 15 years of age by current activity, age and gender, by settlements, Croatian Bureau of Statistics.

The mostly elderly island population has become almost wholly dependent on Zadar health and social care. Most islands have general practitioners, but only a few have dentists, and there are homes for the elderly only in Preko on Ugljan and Sali on Dugi Otok. Most islands have no healthcare professionals at all, so primary and specialist healthcare services are only available in Zadar.

Table 6.3. Age Profile of the Population of the Unbridged Zadar Islands, 1971, 1991 and 2011

			Age Group (%)		
Island	*Year*	*Total Population*	*0–14*	*15–64*	*65+*
Dugi Otok	1971	3.907	18.7	64.5	16.8
	1991	2.719	13.6	64.7	21.7
	2011	1.655	12.6	52.3	35.1
Ist	1971	410	16.6	68.3	15.1
	1991	229	10.5	59.0	30.6
	2011	182	0.5	46.2	53.3
Iž	1971	1.299	12.4	61.7	25.9
	1991	641	8.3	54.3	37.4
	2011	615	8.6	50.7	40.7
Molat	1971	543	12.7	64.3	23.0
	1991	204	3.4	40.2	56.4
	2011	197	1.5	35.5	62.9
Olib	1971	567	14.1	68.8	17.1
	1991	702	5.1	72.4	22.5
	2011	140	3.6	50.0	46.4
Ošljak	1971	0.0	0.0	0.0	0.0
	1991	38	0.0	52.6	47.4
	2011	29	6.9	58.6	34.5
Pašman	1971	3.701	22.4	61.3	16.3
	1991	3.121	17.8	63.5	18.7
	2011	2.845	17.8	57.8	24.5
Premuda	1971	152	15.8	58.6	25.7
	1991	68	4.4	51.5	44.1
	2011	64	6.3	51.6	42.2
Rava	1971	234	14.1	65.8	20.1
	1991	107	0.0	56.1	43.9
	2011	117	5.1	36.8	58.1
Rivanj	1971	54	11.1	68.5	20.4
	1991	20	0.0	35.0	65.0
	2011	31	0.0	45.2	54.8
Sestrunj	1971	353	19.0	67.4	13.6
	1991	96	5.2	52.1	42.7
	2011	48	4.2	47.9	47.9
Silba	1971	337	15.1	52.8	32.0
	1991	217	17.5	47.0	35.5
	2011	292	7.9	59.6	32.5
Ugljan	1971	9.775	18.0	68.4	13.6
	1991	6.941	16.5	64.6	18.9
	2011	6.049	14.9	58.0	27.2
Vrgada	1971	406	20.2	63.3	16.5
	1991	228	13.6	58.8	27.6
	2011	249	12.0	47.8	40.2

(continued)

Table 6.3. *(continued)*

Island	*Year*	*Total Population*	*Age Group (%)* 0–14	15–64	65+
Zverinac	1971	146	13.7	71.9	14.4
	1991	57	1.8	45.6	52.6
	2011	43	0.0	53.5	46.5
Total	1971	21.884	18.2	65.6	16.2
	1991	15.388	14.7	63.3	22.0
	2011	12.556	13.9	55.7	30.4

Sources: Population Census of Households and Apartments 1971, Gender and Age—Part One; results by settlements and municipalities, volume VIII, State Bureau of Statistics, Belgrade, 1973: Population Census 1991; Population by Gender and Age, Croatian Bureau of Statistics, doc. 882, Zagreb, January 1994; Population Census, Households and Apartments 2011, population by age and gender, by settlements, Croatian Bureau of Statistics. Accessed November 9, 2018. http://www.dzs.hr.

Managing the island space, or rather moderating the socioeconomic development of the Zadar islands, has been hindered by administrative-territorial fragmentation. The island of Pag is divided between two counties; there are three local self-government units (LSU) municipalities on Ugljan and two on Pašman, while Vir is an LSU in itself. The island of Dugi Otok forms an LSU with the neighbouring island of Zverinac. The islands of Sestrunj, Rivanj and Ošljak are parts of the municipality of Preko on the island of Ugljan. The island of Vrgada belongs to the municipality of Pakoštane on the mainland, while the island of Babac, which until recently did not count statistically as a settlement, although it has inhabitants, is part of the mainland municipality of Sv. Filip i Jakov. The other islands in the urban archipelago are part of the city of Zadar. The spatial development of the islands is therefore planned and implemented pursuant to 10 municipal or city spatial plans, and socioeconomic life is directed according to 10 development strategies. According to current legislation on spatial planning, the physical plans of cities and municipalities must be aligned with the county physical plan, which must be aligned with the master plan at the national level. On the other hand, the legislation on regional development, socioeconomic plans and the programmes of cities and municipalities do not need to be aligned with the county development strategy or other strategic documents at the county level.

In such a small, geographically dispersed area, in such administrative and planning conditions, it is challenging to achieve developmental complementarity, primarily as the current administrative organisation focuses on maintaining the former functional connection between Zadar and the islands. Local specificities and interests have been completely unincorporated, and this particulate approach has dominated reasoned, scientifically based considerations. For a person living on Sestrunj or Rivanj to get to the municipal cen-

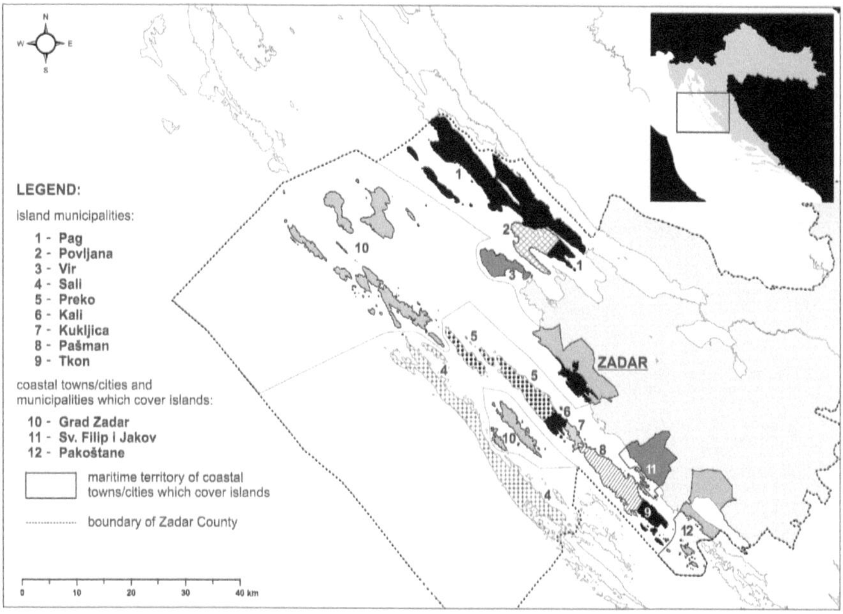

Figure 6.2. Administrative-Territorial Organisation of the Zadar Islands

tre in Preko on Ugljan to deal with an administrative task, they would have to travel by boat to Zadar and then from Zadar to Preko. The journey home would take the same route, since there are no other alternatives, meaning the individual would spend an entire day travelling there and back, although there are only a few hundred metres between Ugljan and Rivanj, and perhaps twice that distance between Ugljan and Sestrunj. The people of Zverinac can reach Dugi Otok, the island housing their municipal centre, by a regular boat service; however, if they do not have their own means of transportation on Dugi Otok, they will need to adjust their journey to correspond with days when the buses refuel at Zaglav near Sali on the southeast end of the island. Otherwise, the bus only serves the villages in the northwest part of the island.

These issues have created a lack of administrative coordination in the Zadar urban archipelago and shown that certain legal and sublegal acts are unaligned or inappropriate. The decision on ranking 556 Croatian self-government units (LAU 1 in the NUTS classification) according to their development level (OG 132/2017) rendered by the ministry in charge for regional development is a good example. It ranks the city of Zadar relatively highly, so that the actual low development level of the islands remains masked beneath the estimated development level of the city as a whole. Acknowledged as suburbs but not as islands, Premuda, Silba, Olib, Molat, Ist and Iž are assigned the development rank of the city of Zadar, which makes them ineligible to apply for various kinds of financial support.

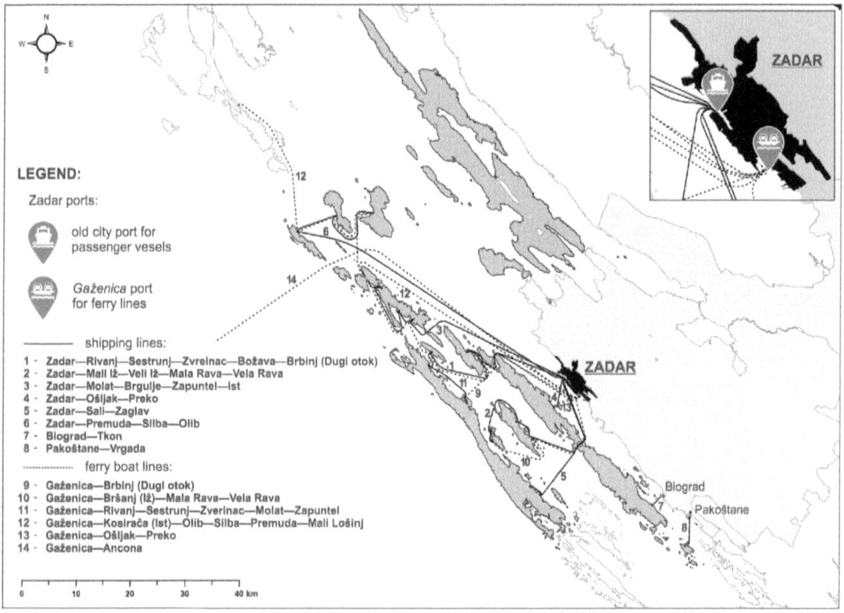

Figure 6.3. Transport Connections Between the Zadar Islands and the Mainland

CONCLUSION

A historical overview reveals that most of the small, inhabited Croatian islands have established functional systems with towns or cities on the neighbouring mainland or larger islands. Smaller islands have always been the rural periphery and thus socioeconomically linked to the towns to which they gravitated. Urban–rural links between the islands and city are still obvious today on the Zadar islands. Scattered and mostly small, the Zadar islands have never had enough resources to develop independent towns and have gradually become incorporated in the Zadar urban network. Zadar was the reference point from where island socioeconomic activities were planned and the relevant administrative decisions made. Human and other resources were drawn from the islands. The islanders, particularly those who worked land owned by wealthy Zadar citizens, had to supply the city with olive oil, wine, fish, lime, pottery, stone slabs and wood for fuel for centuries. The city depended on them, particularly during the Venetian–Croatian–Ottoman conflicts, when most of the Zadar coastal area was under the Ottoman Empire, cut off from direct urban influences and with no supply route from the mainland rural area. At the same time, the islanders turned to the market in Zadar for products that they could not locally produce.

In the modern era, the island economy is based mostly on tourism, while agriculture has largely been abandoned. Today, the islanders travel to the mainland for most agricultural products. Island dependence on the city has steadily grown noticeably, as the number of island inhabitants continues to fall, leaving those who are increasingly elderly, biologically nonviable and economically less active. It seems that the small Zadar islands will soon become occasionally inhabited places. This will jeopardise traditional island culture and reduce two-way urban–rural connections primarily for rest and recreation. This perspective is a challenge to the implementers of island spatial-economic development policies. To date, in spite of appropriate infrastructure investments and the introduction of reasonably good shipping lines, the challenge has not been taken up.

At the global level, there is a lack of studies regarding the specific socio-geographic development of near islands and the creation and formation of this type of urban archipelago. This research is the result of initial investigations that, if similar correlations are acknowledged, may form the basis for future research that would include near islands elsewhere in the world and indicate the complexity of mutual relations between the islands and the neighbouring mainlands, and the relative concept of remoteness in relation to the island space. Remoteness, even in terms of near islands, cannot be seen merely in terms of geographic position in relation to the mainland but needs to include such factors as traffic and functional connectivity, and perhaps the experiences of the islanders themselves. This is the only way to determine whether an island can be classified as remote or near. Given the example of the Zadar islands, which fall into the category of near islands, there is an obvious difference between those on the edge of the archipelago and those located close to the mainland. The outlying islands are characterised by depopulation, ageing and economic regression, while the nearer islands enjoy more intense, complex links with the mainland urban centre and thus experience more stable demographic development and economic vitality.

REFERENCES

Census of Population, Households and Dwellings 2011, Croatian Bureau of Statistics, Zagreb, www.dzs.hr.

Čuka, Anica. 2013. "Razvoj poljoprivrede i promjene krajolika sjeverozapadnog dijela Dugoga otoka." In *Veli Rat*, edited by Ante Uglešić and Josip Faričić, 541–68. Zadar: Sveučilište u Zadru.

Čuka, Anica. 2008. "Geografski uvjeti razvoja poljoprivrede otoka Rave od 18. st." In *Otok Rava*, edited by Josip Faričić, 495–512. Zadar: Sveučilište u Zadru, Razred za prirodne znanosti HAZU, Matica hrvatska—Zadar, Hrvatsko geografsko društvo—Zadar.

Čuka, Anica. 2006. "Utjecaj litoralizacije na demogeografski razvoj Dugog otoka." *Geoadria* 11, no. 1: 63–92. https://doi.org/10.15291/geoadria.104.

Čuka, Anica, Lena Mirošević, Josip Faričić, and Vera Graovac Matassi. 2017. "Phylloxera Revisited: The Spread of Grapevine Disease in Dalmatia and Its Influence on Socioeconomic Development and Agricultural Landscape." *Annales, Series Historia et Sociologia* 27 (1): 101–18. https://doi.org/10.19233/ASHS.2017.09.

Faričić, Josip. 2012. *Geografija sjevernodalmatinskih otoka*. Zagreb: Školska knjiga.

Faričić, Josip. 2003. "Otok Pag na starim kartografskim prikazima." *Geoadria* 8, no. 1: 47–126. https://doi.org/10.15291/geoadria.95.

Faričić, Josip, and Lena Mirošević. 2014. "Artificial Peninsulas and Pseudo-Islands of Croatia." *Annales – Series historia et sociologia* 24, no. 1: 113–28.

Faričić, Josip, Anica Čuka, and Vlatka Colić. 2010. "Poljoprivreda i razvoj ruralnoga krajolika Ista i *Škarde*." In *Otoci Ist i* Škarda, edited by Josip Faričić, 573–98. Zadar: Sveučilište u Zadru, Zavod za prostorno uređenje Zadarske županije, Matica hrvatska—Ogranak u Zadar i Hrvatsko geografsko društvo—Zadar.

Faričić, Josip, Vera Graovac, and Anica Čuka. 2010. "Mali hrvatski otoci—radno-rezidencijalni prostor i/ili prostor odmora i rekreacije." *Geoadria* 15, no. 1: 145–85. https://doi.org/10.15291/geoadria.548.

Graovac, Vera. 2004. "Islands on the Verge of Extinction: The Example of Zadar Islands, Croatia." *Geoadria* 9, no. 2: 183–210. https://doi.org/10.15291/geoadria.136.

Korenčić, Mirko. 1979. *Naselja i stanovništvo SR Hrvatske 1857–1971.—Djela JAZU, knjiga 54*. Zagreb: JAZU.

Population Census of Households and Apartments 1971, State Bureau of Statistics, Belgrade.

Population Census 1991, Croatian Bureau of Statistics, Zagreb, www.dzs.hr.

Ratter, Beate M. W. 2018. *Geography of Small Islands: Outposts of Globalisation*. Cham: Springer.

Royle, Stephen A. 2001. *A Geography of Islands: Small Island Insularity*. Routledge Studies in Human Geography. London and New York: Routledge.

Rubić, Ivo. 1952. *Naši otoci na Jadranu*. Split: Izdanje Odbora za proslavu desetogodišnjice Mornarice.

Settlements and Population of the Republic of Croatia, 1857–2001, Croatian Bureau of Statistics, Zagreb, CD ROM.

Starc, Nenad, Marija Kaštelan-Macan, and Silvana Ćurlin, eds. 1997. *Nacionalni program razvitka otoka*. Zagreb: Ministarstvo obnove i razvitka.

Zadar State Archives (Državni arhiv u Zadru). 1759a. Fond obitelji Lantana, kut. 2, sign. 45, ozn. 8, anagrafska tablica iz 1759. (HR-Zadar State Archives D-359.)

Zadar State Archives (Državni arhiv u Zadru). 1759b. Fond obitelji Lantana, kut. 1, sign. 43, anagrafska tablica iz 1759. (HR-Zadar State Archives D-359.)

Zadar State Archives (Državni arhiv u Zadru). 1759c. Fond obitelji Lantana, kut. 2, sign. 45, ozn. 20, anagrafska tablica 1759. (HR-Zadar State Archives D-359.)

Chapter Seven

Near, Far or Beyond?

Migration from the Dalmatian Islands of Olib and Silba

Tomislav Oroz

The north Dalmatian islands of Olib and Silba belong to the Zadar archipelago presented in the previous chapter. To a mainland outsider navigating according to mainland experience, they may look like near islands. As the crow flies, and nautically speaking, their distance from the mainland supports this assumption (see Figure 6.1 in Chapter 6); however, to islanders, the notion of distance does not necessarily correspond to the cartographic representation of space. Islanders do not measure distance from the mainland in kilometres or miles, but in time spent on ferries, which, according to some, sail on their lifeline routes (Mendas 2015). Whether a ticket-seller for the ferry line, an islander heading to Zadar or a tourist for whom the island has become rather cramped, requests for information are answered in terms of hours; everyone focuses on departure and arrival times, and the weather is important in relation to whether it will speed up or slow down the journey.

For Silba and Olib, the "distant proximity" (so near but so far) paradox is even sharper, since these near islands are more than three hours away from Zadar by ferry, with occasional dockings on neighbouring islands. The paradox is not the only anomaly that characterises life on Silba and Olib. In tourist adverts, they are often portrayed as oases of unspoilt nature, belonging to Zadar in the administrative sense but physically separated from it. Managing the island space is often depicted as a difficult task, since there are only a small number of local boards, and they must gain the approval of bureaucrats on the mainland for every initiative. Any attempts to overcome the almost colonial attitude of the city toward the islands induces the populations of Silba and Olib to claim their rights through unwritten rules and practices that are still in their own hands.

The issues of belonging to the island, the questions of identity and the right to be considered as islander are entangled in a complex web of historical, social and cultural contexts. After conducting several years of ethnographic field research on Olib and Silba, I will attempt here to consider the relationship between the practice of island belonging and being an islander through a theoretical reflection on migration and island studies. In the transnational paradigm within migration studies the experiences of migrants are observed as multiple, layered and complex, and build up their own epistemology, transcending the limits of methodological nationalism and the nation-state container theory of society. This has enhanced the understanding of migrational realities as transborder and transorganisational social phenomena (Levitt and Glick Schiller 2004); however, theoretical reflection inspired by the paradigm of transnationalism goes beyond the specific example of Olib and Silba because it helps to stir the debate about the usually uncritically perceived naturalised relationship between islanders and their island. In the light of nissology, where archipelago metaphors exceed the usual antonym association with islands (Depraetere 2008), the questions of the entangled island experience seem not only relevant, but also crucial in terms of understanding epistemology of island studies. In other words, understanding island identities takes us beyond mere physical presence on the island, into the realm of social and cultural coordinates of what constitutes the sense of belonging to the island. This, however, does not necessarily imply the need of conducting research on the island.

Early 20th-century anthropological research inaugurated by doyens of the discipline like Bronislaw Malinowski (the Trobriand Islands) and Margaret Mead (Samoa) focused on the island communities. Island-oriented field research enabled the paradigm of the laboratory of culture, where the misconception of culture perceived as static was explained in relation to the physical distance from the mainland (see Gillis 2004, 116–20). The notion of belonging in regard to Silba and Olib, however, avoids the trap of essentialisation of island culture. James Clifford's remark that the notion of distance, rather than mutual connection, has contributed to the naturalisation and Othering of the cultures researched (Clifford 1997) warns us that to rethink critically the island cultures in all of their complexity, one needs to abandon the stereotypical, hermetical assumptions generated on the mainland. In a time of deterritorialisation, as Arjun Appadurai (1996) notes, when the notions of culture, space and identity have been uprooted from their traditional setting, such metaphors as "island paradise," "the Mediterranean as it once was" and "untouched, virgin space" cannot serve as the starting points that will help us to illuminate the aporia of entangled island experience (51). Metaphors of islands and island communities that nurture the idea of isolation are the legacy of islomania, which John Gillis defines as a specific mainland perspective in considering the island concept, a sort of *long durée* of the Western imagination that combines the ideas of isolation, connectivity, separation, the past and the future (Gillis 2004).

In the land of a thousand islands, as Croatia likes to flatter itself, thinking about the islands often assumes taking an uncritically mainland perspective rather than the islands one; therefore, the problems associated with identity have often been posed by limiting and stereotypical assumptions depicted from the mainland that contributed to romanticizing about island communities. Questions about the everyday practices of the islanders and their identity cannot be answered if the island experience is viewed only from the epistemological safety of the mainland. To understand the complex migration experience of Silba and Olib islanders, it is necessary to use the methodology of ethnographic research to comprehend the interweaving of everyday practices and social relations that depict the islanders' ways of life, along with the memories and nostalgia stirred by the island narratives and identity strategies thus pointing out to diverse ways of belonging (Levitt and Glick Schiller 2004).

Although Olib and Silba are separated by less than four miles, their proximity is diametrically opposed to the variety of their historical, social and cultural facets. Early on, Silba, which has been largely known as an island of seafarers and captains, based its economy on the capital that came to the island from abroad. Work migration thus became an integral part of island life, a sort of island brand, whereas the sense of island belonging was marked with continuous translocal practices involving contacts with places and cultures beyond the island. Changes in the island economy, which by the beginning of the 20th century was grounded in livestock farming and seafaring, occurred with the tourism. Silba was "discovered," particularly during the 1970s, when it became a safe haven for naturists and those with alternative lifestyles, for instance, musicians and hippy groups. The reverberations and effects of these changes have endured until today and continue to draw artists and performers to Silba, where they find inspiration.

On the other hand, the neighbouring island of Olib based its economy right up until the late 19th century on agriculture and livestock farming. Transoceanic emigration began when the economy began to stagnate as a result of Peronospora, followed by phylloxera and the introduction of the Wine Clause in the 1891 trade agreement between Italy and Austro-Hungary, which favoured Italian wines. Redefined experience of island life marked by migration thus became the leitmotif of Olib. In less than 100 years, the migration experience of the islanders changed the face of the island. Today these migration footprints are apparent at every step—from English words that found their way into the local dialect, through new houses built with American porches, all the way to daily routines adopted in the United States and practised on Olib.

By nature, migration connects the islands with other places (King 2009). In the cases of Silba and Olib, migration is an inevitable reference in understanding the past and present but also the creation of identity based on perspectives and experiences acquired beyond the island—in the diaspora. Diaspora as such is crucial in understanding of identity as a way of belonging because as a concept it enables us to reflect on the process of making a home for those migrants whose life is marked by constant movements. Bearing in mind constant identity transformation due to intense entanglement in movements beyond island boundaries, the diaspora identity involves what Avtar Brah calls *multi-locationality*, or the process of diasporic identity creation that ignores imposed territorial, cultural, and psychic boundaries (Brah 1996).

Both Olib and Silba have based their relations at the symbolic level on historically inherited, proverbial animosities, through which one can reflect on their inherited class, economic, cultural and identity particularities. Here, we need to pay attention to the dynamics of the archipelago experiences of Silba and Olib in a synchronistic and diachronistic perspective but also bear in mind the question of how the islanders' ways of being and belonging have been formed in relation to the historical, social, economic and cultural contexts. The answers to these questions can be considered in the light of a theoretical reflection on the transnational paradigm of migration studies that shed light on the ambivalence of belonging, while at the same time not ignoring the practical experience of proximity and distance from the perspective of the islanders themselves. As far as Silba and Olib are concerned, these experiences are undoubtedly linked to a mainland perspective that still continues to determine the lives of the islanders; therefore, Godfrey Baldacchino's warning about the seductiveness of reductionism in researching the islands should be understood as a milestone marking a step forward from stereotypical assumptions toward factors that can be confirmed by island experience (Baldacchino 2008).

The life of the Mediterranean islands, with their internal, restricted, intimate existence and, at the same time, the important role they have on the stage of world history (Braudel 1972), is relevant in understanding the experiences of Silba and Olib. Alongside the obvious, external face of the islands, dominated by mainland perceptions of small island communities, another aspect must be acknowledged that transcends the limits of physical barriers and illuminates the dynamics and adaptability of the island communities. Since the temporary and permanent departures of the island population have essentially determined their sense of belonging to Silba and Olib, the diversity of transnational practices must lead us to examine the symbolic borders that constitute the island space. What does it mean to belong to these islands, to call oneself an islander or to stake a claim to the islands when migration experiences have erased the traditional criteria for islandness?

SILBA: FROM A SEAFARING ISLAND TO AN INSPIRATION FOR ARTISTS

Silba is the most northerly of the Zadar islands, located 29 nautical miles from the city of Zadar. With a surface area of 15 km^2, it belongs to the group of small Adriatic islands (Magaš and Brkić-Vejmelka 2013). Two associations dominate public discourse about Silba. To nonislanders and tourists, Silba is one of those rare Croatian islands where road traffic is banned. So Silba is often portrayed as a natural paradise, with no crowds or traffic jams, surrounded by the crystal-clear sea—a safe haven for artists and those who lean toward an alternative lifestyle. This public image dates back to the 1970s, when members of the alternative music scene recognised its potential for creative inspiration, artistic freedom and nudist bathing.

The island paradise scene was not just the result of the crazy 1970s or contemporary tourist propaganda, but was actively promoted by the islanders, who associated the notion of paradise with the economic prosperity of the island in past centuries. The proverbial memory of Silba's golden age, when it was "paradise on earth" (Jurić 1910, 13), is a consequence of the island's particular maritime orientation. In this context, the self-imposed vision of Silba as a captains' island points to not only the dominant economic branch, but also the symbolic aspect of identity in which highlighting the captains' profession often reflected on family honour and internal social and class stratifications. In a cultural analysis, these apparently different perceptions prove motivational, allowing an understanding of the complexity of living on the island. The island boundaries that Peter Hay claims "invite transgression" and also "inspire restlessness," and so "demand to be breached" (2006, 23), in the case of Silba, figure as leitmotif of the island's past and present. A wide range of migration practices, from work migration, then transoceanic, to modern lifestyle migration, have marked the island's past and present.

Although the vision of Silba as a captains' island no longer reflects the economic reality today, a newcomer cannot miss a monument honouring the most esteemed aspect of the island's past. The totemic sculpture *Sails*, by sculptor Šime Vulas (Srhoj 2014), located on the dock, is in memory of the seafaring people of Silba, who made the island into a sort of "maritime nest" (Bošnjak 1961, 35). As a "visual symbol of Silba" (Domljan 2013, 236), along with the still-remarkable captains' houses built in the late 18th century (Starešina 1971), it functions as a landmark that makes the maritime orientation of the island visually present.

During field research, many of islanders proudly presented us with traces of Silba's maritime tradition in their own houses. Almost every conversation began with them showing us their household relics, for example, family photographs, portraits of captain ancestors or pieces of ship's tackle illustrating the maritime tradition that was affirmed historically in the late 16th century (Bošnjak 1961). At that time, the Venetian Republic was building up a system of maritime routes along the eastern Adriatic coast. Venice considered islands most convenient shelters, almost like a part of the fleet (Braudel 1972); therefore, the Venetian word for sailing (*costeggiare*) actually meant following the coastline—sailing by day and using the natural shelter of island harbours by night (Braudel 1972). Small vessels on the way from Venice to Zadar passed close to Silba (Fortis [1774] 2004, 6).

In the context of intense communications, which were the most dominant feature of the early modern Mediterranean (Braudel 1972), Silba built its economic prosperity by linking up with trading networks. This was most obvious in the 18th century, when sailors and captains from Silba sailed the Mediterranean and European Atlantic coastlines (Vigato 2013). The enduring connection with life at sea (Bošnjak 1961) was evident in the creation of a merchant fleet, owned by certain families, which traded up and down the Mediterranean (Bošnjak 1961). The Napoleonic Wars, the collapse of the Venetian Republic and the modernisation of shipping traffic during the 19th century led to economic stagnation on Silba, a tendency reflected in demographic data. Attempts to revitalise the island's maritime orientation in the 1820s, when shipping was renewed throughout the Adriatic (Starešina 1963), eventually came to nothing in the face of modernisation processes and the arrival of steamships (Mirošević and Faričić 2013).

The external, public face of Silba, with its maritime heritage and captains' tradition, representing the most prominent aspect of the island's history, is often referred to today by the people of Silba as ground for identity strategies of the island's social community. The genre features of oral tradition representing the past bear witness to the symbolic attitude of the community toward the past, pointing to the interweaving of historical references, social changes and narration inspired by symbolism (Agoston-Nikolova 1989). Early 20th-century works highlight the maritime past in descriptions of particular items of interest on Silba (Jurić), and the island crest bears a sail as a symbol of the island; however, this insistence on family honour, linked to the captains' profession and connections with the island economy, can create a rather monolithic perception based on only one historical period. An examination of economic practices and trades, which oral tradition tends to gloss over, and which seem less attractive from today's perspective, may open the way to a better understanding of symbolic aspects of internal valorisation through which the complexity of social and class relations can be assessed.

The island is also linked with fishing, which is often described in the literature as the precursor of shipping (Cvitanović 1962). Data acquired during anthropological research in the 1980s mentions fishing as an economic activity on Silba but on the margins of more prestigious sailing, which was well-developed (Tomljenović 1983). An analysis of interviews conducted during ethnographic research showed that in the islanders' own perception of their identity, they tended to exclude agriculture, which they thought was characteristic of neighbouring Olib. Attributing agricultural traditions to Olib (the Olib crest bears a grapevine) is often correlated with proverbial interisland animosities and an emphasis on the urban character of Silba associated with elitist maritime connotations. Since the people of Olib often worked on Silba as hired hands, their social status in the eyes of the people of Silba was not to be envied and was associated with hard physical labour. Ethnographic notes from the 1980s confirm that the people of Silba had not thought of themselves as "peasants" for hundreds of years (Nikočević 1983).

Agriculture was connected with not only the island landscape of Silba, but also the wider archipelago, particularly the islands of Pašman and Pag, where estates owned by people from Silba were leased to the villagers, tenants in the *colonat* system (Starešina 1971). The fact that the people of Silba emphasised sailing and suppressed agriculture in their narrations about their past indicates that they perceived their history as the age of the greatest economic progress on the island, which is linked to the capital brought in by maritime affairs. Viticulture and the negative associations with it in the second half of the 19th century were accompanied by the islanders emigrating and the impossible challenge of recovering the economic strength of Silba's golden age.

The opposing urban and rural facets of the island show that the complexity of internal social stratification and the perception of the values were associated and popularly justified with the class belonging, cultural background, and even atavistic racial characteristics. According to some of the authors, these two faces of the same island associated with the seafaring and farming were argument for the concept of duality of island culture. According to Starešina (1963), since the 17th century, the typical people of Silba have personified both labourers connected to the soil and shipowner captains or sailors connected with the wider seas. Recognising that this duality is the result of patriarchy and conservatism that grew out of a rudimentary economy based on the *colonat* relationship and the more advanced ideas shipping brought to the island, Starešina considers native characteristics and atavistic perspectives, and builds the duality hypothesis.

According to Starešina, this duality is evident to each person from Silba and linked with the patriarchy supported by the women or the cosmopolitan spirit displayed by the men. He explains this situation in terms of the criteria of residence or translocation, thus introducing a division that says more about the internal, unwritten rules of (self) perception in relation to social values than an analytically intriguing perspective from which to understand the dynamics of the island community. On the other hand, this division indicates urban, bourgeois and rural, agricultural identity points of reference that are linked discursively with the value system, hence describing the island space in essentialist terms; therefore, textual reconstruction of identity is promoted through the dramatic dual-culture concept that articulates the (un)desirable criteria of belonging to the island. This discursive dramatisation of the island history and society has symbolic consequences that are present in not only the more frequent naturalisation of bonds between the people and the land, but also the description of livestock farming, which is characterised as primitive (Vigato 2013). Thus, the question arises as to why seafaring is given a higher place in internal value ratings, while agriculture is marginalised, and how this has influenced the dominant image of Silba's past. Is the island identity rooted in the physical island space or entangled in a symbolic social network, built on class, economic and historical discourses beyond the given limits of the island? As a consequence, does it determine the translocation of the Silban identity?

The answer to these questions must take into consideration the historical context, which is often stressed as the primary argument for the current state of affairs on Silba in respondents' accounts. When asked about life on the island today or problems relating to connections with the mainland, islanders often began their answers by referring to the past, which they claim has determined the current situation. One of the key reference points in their statements was the year 1852, when the island land was redeemed from a man named Raguzin from the island of Lošinj. Before that, the island had been the property of various Venetian families for centuries. In the 15th century, aristocrat Šimun Detrešić was mentioned as the leaseholder. The last one was Benedikt Soppe. During the era of the Venetian Republic, it was decided that all state goods should be sold, so Ivan Krstitelj Soppe bought Silba with his son, Frano, on 20 July 1639, for 12,350 ducats. In 1778, the Contarinis were still the proprietors, but in 1808, the island passed into the ownership of the Morosinis, a Venetian patrician family (Starešina 1973). At first, the people of Silba paid annual rent in kind (Vigato 2013), but beginning in 1770, they paid in cash, which was a direct consequence of the boom in currency due to the development of shipping (Starešina 1971).

When the island was bought from its owner, the *colonat* system came to an end, and this had direct effects for the islanders' social status. At last, islanders finally owned the land they lived on. Unfortunately, this positive step forward was blighted by the arrival of phylloxera. Silba was the first of the Zadar islands to be affected (Mirošević and Faričić 2013). By analysing demographic trends on Silba from 1857 to 2011, Mirošević and Faričić have indicated the correlation between demographic and economic decline in the second half of the 19th century, which resulted in the first strong waves of emigration from the island (Mirošević and Faričić 2013). While the nearby islands experienced demographic growth between 1857 and 1910, Silba lost 40.4% of its population (Mirošević and Faričić 2013). Negative demographic trends continued after the First World War and throughout most of the 20th century. Administrative reorganisation and political changes after the Second World War set the socioeconomic situation on the island even further back, so that Silba found itself on a "double periphery"—in relation to the national centre and regional political-economical centres (Mirošević and Faričić 2013).

The experience of island life characterised by migration manifested itself in various ways, depending on the social, political and economic circumstances in the wider environment. At the turn of the century, people moved to nearby Zadar or Trieste, or emigrated across the Atlantic to the United States and Argentina (Rudan, Sulojdžić, Chaventré and Borot 1987). In parallel with depopulation, the first tourist accommodation appeared on Silba (Brkić-Vejmelka 2013). Although Silba's position on the tourism market dates back to the 19th century, it was only in the 1930s that tourism developed into something more than a mere curiosity. In 1938, Silba, Pag and Zlarin were the only north Dalmatian island settlements to be given the status of tourist resorts. Although statistical data show that the island lost three-quarters of its population in less than a century, some authors see a direct link with the disappearance of seafaring and the rise of viticulture as the main economic activity. During the field research, islanders mostly focused their narrations of the islands' past on that part of the history that helped to maintain the image of bygone fortune and prosperity. Here lies the key to understanding the relationship between being an islander and belonging to the island in the contemporary context. Do the reasons why the migration does not seem as traumatic to them as it does to those living on the neighbouring island of Olib lie in the practices of constantly crossing island borders?

The present impression of Silba as a haven for the alternative art scene began in the 1970s, when artists, musicians and performers established a colony in the then-naturist camp, and the island quickly gained the cult status in cultural milieus and artistic circles (Nikačević 2018). At the same time, depopulation was progressing rapidly on the island, as the industrial boom in Zadar was attracting people from the surrounding areas (Mirošević and Faričić 2013). According to contemporary witnesses, the anthological achievements of the Yugoslav New Wave music scene actually happened on Silba. Dobra Voda Beach was particularly inspiring to young musicians who developed a network of acquaintances free from pressure in a creative leisure environment (Nikačević 2018). Even before this, Silba had been promoted by the presence of well-known intellectuals, artists and scientists (Brkić-Vejmelka 2013).

Today, the island is still dominated by sculptures, and a gallery was opened in the village by sculptor Marija Ujević-Galetović. Sculptures of cat women in public parks, sculptures on house fronts and artistic performances have become part and parcel of contemporary Silba.

> For four or five years, Silba was a cult location, a place of freedom. The posthippie fraternity gathered there: people with long hair who believed in the ideals of peace and love. Lots of music. . . . Well, you came there, spent a month or so and let your hair down. We slept in the woods and went to the village for food and water. It was an ideal place for young people to develop spiritually! The beach there is almost Caribbean, with wonderful white sand and a turquoise sea. On the beach, mosquitoes flied and girls ran around. The camp was naturist, so we simply fitted in with that global trend. People were always making music in the bushes, and it was all pretty relaxed as far as social and political supervision was concerned. Particularly in regard to sex, it was pretty immoral in some ways. (Petrinović 2006, 51–52)

During interviews, people often emphasised that the practice of selling plots of island land to nonislanders demonstrated the fact that the people of Silba had lost touch with their island as the symbolic source of their identity. At the same time, they regarded the more circumspect sale of land on Olib as a positive sign and ascribed it to the social cohesion of the Olib community, which had managed to survive in spite of intense emigration; however, the practice of selling land should not be interpreted as a loss of symbolic ties with the island, or even a deficient sense of belonging to the island, but the result of the historical legacy in which seafaring played an important role. In essence, seafaring meant venturing beyond the island boundaries and proved that the island was not the only point of reference in identity strategies. The marginal position of agriculture, and the negative social and class connotations associated with it, revealed that the process of forming the Silban identity actually depended on locations beyond the islands. To understand the translocal character of the Silban identity, the concept of residence is indicative; many Silbans have moved to Zadar but retained Silba as their official address. Apart from the tax benefits this brings (which are often mentioned as the main incentive), interviews revealed that many islanders do not see this simply as a means of personal gain, but as a way of expressing their connectedness with the island. Hosting artists in the 1970s contributed to the image of a community that was open to new cultural phenomena from beyond the island, phenomena that enriched its existing cosmopolitan connotations.

Following leads about the positive, creative atmosphere that artistic circles have associated with Silba, modern newcomers, inspired by the narrative of the island open to alternatives, continue to find a temporary home there.

Interviews show that for some of them, Silba is a place where the popular image of islands and the Mediterranean in general mesh with the specific features of this island. The buzzwords used are "discovery," "paradise island" and "virgin nature," which can be applied to many islands, of course, but the experiences of modern lifestyle migrants are particularly positive because of the way they are accepted by the local community. They attribute this special social atmosphere to the island's "positive vibe," which implies freedom, creativity and nurturing artistic individuality. What is more, at the time of the research, Silba's local board comprised some people who were not from the island but who had contributed to improvements. Compared with circumstances on other Croatian islands, where unwritten traditional rules hinder initiatives and thwart the efforts of outsiders, Silba's town council, led by women, proves that the island identity and belonging are being built within the complex social webs characterised by diverse cultural practices. These practices where translocal and autochthonous, modern, and traditional are joined together and impugned with the sense of belonging help to create the uniqueness of islands' identities so often evoked in the narratives of the islanders. Whether by participating in the artistic life of the community through organising exhibitions, staging performances for the islanders, arranging communal dinners and social events or simply going for a morning coffee with the locals, the Silban identity persists in transcending the symbolic and physical barriers that enable entanglement in the wider social scene.

AN OLIB ODYSSEY: FROM AN ISLAND HOME TO TRANSNATIONAL SOCIAL NETWORKS

Most visitors to the island of Olib quickly gain the impression that nothing much happens there. On this distant island on the margins of the tourism market, life goes on at a different pace. In contrast to the bustling tourist towns on the mainland, Olib is a sleepy sort of place. There is one pub where you can listen to classical blues, one restaurant that is closed for most of the year and a few benches in the harbour where the locals eagerly await the next boat; however, if we get past these first impressions, we will discover dynamic cultural processes that reveal another side to the island, devoid of mainland stereotypes. If we spend time with the islanders and listen to their conversations, we will soon unmask that supposed lack of activity and discover that the apparently slow pace of life is also deceptive. The people of Olib also have plenty to say about transoceanic emigration and dramatic memories of leaving the island. Leaving the island was not only a traumatic moment in the lives of individuals, but also had significant consequences for the entire island community.

Statistics show that between 1948 and 2001, the island lost 83.9% of its population (Faričić and Magaš 2002). Beneath the surface of statistical indicators, the emigration in the case of Olib resulted in separating families across the Atlantic, while longing to return home reveals the complex social experience of a tiny community that exchanged its narrow island lanes for the bustling streets of New York. The class, economic, social and cultural changes they experienced in the New World largely redefined their lives, and also changed their sense of belonging to the island. The island's present, migrant past and uncertain future are summed up in a terse retort—*the island is sinking*.

The current situation, when most of the population is elderly, following decades of demographic decline, is not a geographic reality but a social, cultural construct that reflects how the islanders see themselves. The "sinking" island is an ongoing point of reference to how they think of their island identity, located somewhere between the glorious past and the present marked by migration. Their ambivalence in relation to the island experience is a reminder that the island space and belonging to it must be considered in the context of the islanders' everyday lives, rather than in artificially imposed, culturally petrified notions of culture. Understanding the multiple perspectives of the migrants means distancing oneself from focusing methodologically on the island's physical space and turning instead to the transnational migrant space that transcends spatial, political and cultural limits (Glick Schiller, Basch and Blanc-Szanton 1992). Thus, ways of belonging to the island can be discerned in a wide arc, from discursive constructions of Olib as a place to be yearned after through narrative depictions of its past, the practical experiences of the Olib community on American soil and right up to the returnee syndrome and cultural transfers in the island space.

James Clifford (1997) notes that it is not only people who travel, but also their cultures. Movement, therefore, is a constitutive action that forms centres, regions and territories that had no inherent meaning before the movement began (Clifford 1997). The travels and movements of the people of Olib, which began from their island, demand that such concepts as "local" and "belonging" are not seen exclusively or hermetically through the prism of sedentariness of island culture, but as experiences constituted by multiple "belongings," beyond the fixed borders that give significance to the concepts of "near" and "far." Examining and understanding the identity of the people of Olib, who are undoubtedly linked through migration to a space far beyond their island, means understanding from the island perspective and through the islanders' lens their everyday lives, transformed by transnational experiences, and their narratives inspired by the island's past, long before the age of migrants.

Migrations from Olib were part of wider socioeconomic transformations in the eastern Adriatic. According to demographer Ivan Lajić, the main reasons for emigrating from the Dalmatian islands were agrarian overcrowding, the crisis in viticulture, the Wine Clause, debt on the part of the peasants and such factors as the collapse of shipbuilding and the seafaring economy (Lajić 1992). In the late 19th and early 20th centuries, a large number of people from Olib emigrated across the oceans, particularly to the United States. Some authors, like Ivin, argue that there were four periods of intense migration from Olib. The first wave was between 1910 and 1914, mostly to Argentina, Australia and the United States. Most of those who emigrated were young men, and this had a marked effect on demographical trends on the island. The second wave started after the First World War, between 1920 and 1925, and the most common destination was the United States. The third wave started in 1935 and lasted until 1940, while the fourth wave started after the Second World War (Ivin 2009). This was the period when phylloxera devastated the vineyards and damaged the island economy.

Natural disasters and adverse circumstances like the Wine Clause forced the people of Olib to emigrate in search of earnings, no matter how high or low (Dragić 2002). Since then, migration has been a thread running through island life, so much so that no family on Olib has remained unaffected by it. The consequences of transoceanic emigration are clearly shown in the statistics. In just a few decades, the population halved. By the 1953 census, there were only 805 inhabitants (Faričić and Magaš 2002), while during this research, there were a little more than a hundred. After the Second World War, another intense wave of emigration began, caused by economic hardship and political circumstances.

Along with the statistical, historical and demographical, there are also cultural and social approaches to migration. The cultural anthropological approach focuses on how migration affects the reality of the people involved in migration processes. Thus, cultural analysis examines cultural exchanges, negotiating, adapting and reinventing lifestyles in dynamic shifts between cultures. Complex and multilayered, culture is the marker of the individual or collective identity and forms a continual reference point in symbolical demarcation. If migration is understood as a multistage, ongoing process (Čapo Žmegač 2010), "culture in motion," then cultural anthropology focuses on migrant communities and their rules and lifestyles. In the case of Olib, the culture that researchers refer to in terms of Olib's tradition or heritage does not stop at the island's physical boundaries. Rather, a cultural anthropological perspective on migration enables us to understand culture beyond fixed boundaries. Alongside migration, which dominates the narrative of the island's history and because of which it is "sinking," as the islanders say, translocation and transnational experiences form the conceptual framework needed to understand the Olib identity.

The migration history of the islanders is not only the warp and woof of their social memories, but also a visible phenomenon in the island space. A memorial in the part of the village called Igrišće recalls the idealised days before emigration began, when the island was still echoing traditional song. This symbolic spatial marker sits close to abandoned, tumbledown houses that bear witness to the consequences of migration. The islanders' sense of belonging is greatly determined by their perception of the island as it once was and internal rules regarding what it means to be an islander, regardless of their direct experience of living on the island. A visual representation of the island's transnational identity is also found on the Olib crest, which symbolises the émigré Olib community in New York, who gather under the auspices of the Friends of Olib Charity (Oroz and Urem 2015). The dominant image is an apple (a sort of reference to the Big Apple) with a 17th-century fort inside of it. The fort from the crest refers to the real one still present on Olib and built as military fortification to defend the island from pirates during the 17th century; however, locals associate it with early settling of the island in the 15th century by refugees that came from the Cetina river region on the mainland.

Images of Olib as it once was are still important to homesick migrants. They and their descendants show their affection for the island in a variety of ways, from cooking island specialities in their homes and for social events in the Olib Voluntary Society to making sure they have supplies of island olive oil, prosciutto, cheese and other delicacies that can only be sourced from "home." They hold their annual conventions on the 14th of May, a date that links both communities, commemorating the day in 1900 the people of Olib redeemed their island from the Italian landowners. Video links are often used to connect celebrations on both sides of the Atlantic, and communications technology succeeds in uniting the spatially scattered Olib community. In an age when aeroplanes and telephones compress space (Clifford 1997), those living on Olib and those abroad can socialise and communicate via Facebook, in a social space that transcends physical distance.

The benefits of modern communication technology are also evident in the work of the Friends of Olib Charity, which publishes on its Facebook pages photographs of renovated island infrastructure paid for by émigré islanders. The homesickness of those living in the United States fuels their gatherings, where they can practise and narrate "feeling at home." A good example is the music of Cine Pulišić, a musician from Olib with permanent residence in New York, as he himself says. Pulišić recorded an album called *Jednom ću se vratiti* (Someday I'll Be Back). In popular songs like "Someday I'll Be Back," "American Dream," "Sing to Me, My Island," "I Am Sorry," "My Heart Cries," "I'll Be Back, Olib" and "On My Own Island, Olib, My Home," Pulišić evokes nostalgia for his native island on the other side of the world. Some of the songs are in English and some in Croatian, mostly in the Chakavian dialect. Pulišić recalls romanticised images of an Olib left behind by migrants in search of a better life. Memories of childhood, a mother waiting for her son to return, the journey of life and having homes on two continents figure throughout the album. Besides the desire to return, migrants who have actually returned in the last several years have made a cultural impact on the island, transforming the place they once left.

During many years of research on Olib, conversations with returnees would always begin with a review of the current state of affairs on the island. Most of the present population is composed of returnees from the United States who left in the 1960s in search of work. Their decades of absence often included preparations to return in retirement. Among them, many have not decided to return permanently. For them, returning is temporary, and they continue to spend part of each year in the United States. Given the time it takes to reach the mainland or Zagreb, many returnees find travel to and from the United States easier and American soil closer in terms of time. Returning is seen as just one (potentially reversible) step in the migration circuit and not its end (Čapo Žmegač 2010). It is reflected in their concepts of Olib culture, tradition, heritage and redefining the island space. They have tried to enhance this reality by recreating their personal experience of living in the United States and introducing commodities they have become accustomed to during their "temporary absence." How else can we explain the combination of pragmatism acquired in the United States; the Mediterranean siesta; American-style houses in rural Dalmatia; imported food in a country that waxes lyrical about traditional, local food or daily Skype communications with offspring throughout the world paired with poor communication with the nearby mainland?

Perhaps one of Olib's most emblematic symbols today is the buggy—as you arrive, you will see golf carts and all-terrain quad vehicles parked on the ferry dock. You might assume there is a golf course on the island, but you would be wrong. Golf carts were imported from the United States to help elderly people reach parts of the island that once were reachable only with donkeys. Golf carts also evoke the experience of living in the United States, while bumpers stickers reading "I love Olib" unmistakably carry echoes of New York. The island was connected to the mainland electricity network thanks to donations made by islanders living abroad. They also helped rebuild the church and school. Solar panels have been installed on roofs, so the visual appearance of Olib is quite different from most other Croatian islands.

Transposing a lifestyle acquired in the United States to rural Olib is a common, widespread practice; however, the sharp contrast between the romanticised image of the island propagated by the returnees and the island they have remodelled in the American fashion shows that the sense of being an islander and belonging to the island refer to the social space of Olib, rather than the physical island. To avoid "islanding" (Baldacchino 2007, 2) and wrongly representing the islanders' perception of themselves, we should ask what it means to be a part of Olib's social space. Is it merely having a longed-for physical presence on the island or what Avtar Brah calls the homing desire that transcends the physical gap between the home and the place of origin (Brah 1996)?

Answering the question about what it means to belong to Olib depends to a large extent on attitudes toward the land as a symbolic repository for identity creation. A monument erected in 1934 bears witness to the redemption of the land in 1900, and its design and inscription point to the unbreakable connection between two events: settling the island in the 15th century, popularly perceived through the Venetian fort as the visual mark of settling the island, and the date when the land was redeemed from Zadar landlord Dunato Filipovo. Although the Venetian fort that dominates the central part of the island was built in the 17th century to provide defence against pirates, the people of Olib consider it a symbol of the start of the settling of the island.

The foundation of the island economy today, as in the past, is built on agriculture and livestock farming, and, to a lesser extent, seafaring and fishing (Faričić and Magaš 2002). Livestock farming has played an important role in the island economy, and the Olib Sheep Dairy Cooperative, founded on the island in 1933, bears witness to that. The people's connection with the land has been demonstrated in breeding sheep and goats, olives and grapes, and managing the holly oak forests for fuel—the land that represents life in the symbolic imagination of islanders. Since the island has such an agricultural emphasis, the redemption of the land in 1900 represented a drastic change in the way of life, enabling the people to enjoy economic independence that they perceived as freedom. The land they had once leased by redemption became theirs, symbolising a new beginning. As many returnees said, for the first time, it meant they were "on home ground." The inscription on the monument describes the event as defining, while the monument itself is a symbol of freedom. It is ironic that although this act enabled the people of Olib to make a huge stride forward in terms of economic independence, the island economy remained dependent on agriculture, and when crises occurred, the people were forced to leave.

The bond with the land, which meant being rooted in one's homeland and social space, did not break when the islanders moved overseas. The transnational practices of the islanders was grounded in the symbolism of the land, so that expressing belonging to Olib by owning land on the island became much more than mere proprietorship. Land symbolised a bond with one's birthplace, the starting point of one's journeys through life, one's family roots and the Olib community. Even the descendants of migrants who have never been to Olib will not sell their land, because owning it means having a link with the land of their forefathers. In some cases, the land is worth virtually nothing in monetary terms, especially when multiple ownership is involved, but its symbolic value is high, as it represents belonging to the Olib community. Anyone who decides to sell is breaking an unwritten rule, cutting themselves off from their roots and voluntarily excluding themselves from the island community. Roots are a sign of permanent belonging and the hope of returning, wherever one lives.

NEAR, FAR AND BEYOND: CONCLUDING REMARKS ON THE MIGRATION EXPERIENCE OF ISLANDERS

If we step back for a moment from Silba and Olib, and consider them at the level of the archipelago, we will soon see that migration experiences, whether temporary or permanent, are conditioned by the symbolic belonging to the island worlds. Those worlds refer to the islands as physical entities, as well as to their social and cultural coordinates that position the experience of belonging to island worlds. The concepts of borders, the experience of the "near" or "far," from the perspective of the islanders of Silba and Olib, are not abstract or predetermined, but shaped by specific facets of island culture, history and social norms. Any attempt to reconsider the island space and its borders and potential limitations must start from the experiences of the islanders themselves and their notions of identity, locality and belonging.

Research has revealed that migration experiences spur people beyond the limits imposed by geography of the island in transnational spaces beyond a sedentary understanding of culture as a petrified phenomenon. These processes of belonging are complex and indicate the intertwinements of various aspects that form migrant identity. The proximity of islands, however, does not imply the similarities between them in terms of their culture, but rather points to social and cultural complexities of island worlds that reflect on their notions of near, far or beyond. Whether it's about translocal experiences of the islanders from Olib that reinforce their bonds with their native island in transnational space or the experience of Silba islanders for whom the seafaring tradition enabled contact with diverse cultures, the evocation of the archipelago puzzlement when considering islands must take into account high stakes of social, cultural, economic and political contexts that reshape islanders' sense of identity and their notion of near and far.

REFERENCES

Agoston-Nikolova, Elka. 1989. "The Notion of Time in South Slavic Oral Poetry." *Narodna umjetnost* 26: 21–24.

Appadurai, Arjun. 1996. *Modernity at Large: Cultural Dimensions of Globalization.* Minneapolis: University of Minnesota Press.

Baldacchino, Godfrey. 2008. "Studying Islands: On Whose Terms? Some Epistemological and Methodological Challenges to the Pursuit of Island Studies." *Island Studies Journal* 3, no. 1: 37–56.

Baldacchino, Godfrey. 2007. "Introducing a World of Islands." In *A World of Islands: An Island Studies Reader*, edited by Godfrey Baldacchino, 1–29. Charlottetown, Canada, and Luqa, Malta: University of Prince Edward Island, Institute of Island Studies and Agenda Academic.

Bošnjak, Radovan. 1961. "Silba - društveno-geografsko značenje." *Hrvatski geografski glasnik* 23, no. 1: 19–35.

Brah, Avtar. 1996. *Cartographies of Diaspora: Contesting Identities (Gender, Race, Ethnicity).* New York: Routledge.

Braudel, Fernand. 1972. *The Mediterranean and the Mediterranean World in the Age of Philip II.* New York: Harper & Row.

Brkić-Vejmelka, Jadranka. 2013. "Turistička obilježja otoka Silba." In *Otok Silba, prirodno i kulturno blago*, edited by Jasmina Mužinić and Jenő J. Purger, 285–98. Zadar: Sveučilište u Zadru.

Čapo Žmegač, Jasna. 2010. "Return Migration: The Changing Faces and Challenging Facets of a Field of Study." *Ethnologia Balkanica: Journal for Southeast European Anthropology* 14: 227–47.

Clifford, James. 1997. *Routes: Travel and Translation in the Late 20th Century.* Cambridge, MA, and London: Harvard University Press.

Cvitanović, Vladislav. 1962. "Prilog proučavanju pomorstva otoka Silbe i Premude." *Radovi Instituta Jugoslavenske akademije znanosti i umjetnosti u Zadru* 9: 447–74.

Depraetere, Christian. 2008. "The Challenge of Nissology: A Global Outlook on the World Archipelago—Part II: The Global and Scientific Vocation of Nissology." *Island Studies Journal* 3, no. 1: 17–36.

Domljan, Žarko. 2013. "Dva recentna umjetnička ostvarenja." In *Otok Silba, prirodno i kulturno blago*, edited by Jasmina Mužinić and Jenő J. Purger, 234–37. Zadar: Sveučilište u Zadru.

Dragić, Ante. 2002. *Stanovništvo na zadarskim otocima.* Zadar: Otočni sabor.

Faričić, Josip, and Damir Magaš. 2002. "Problemi suvremene socio-geografske preobrazbe otoka Oliba." *Geoadria* 7, no. 2: 35–62.

Fortis, Alberto. [1774] 2004. *Put po Dalmaciji*. Split: Marjan tisak.

Gillis, John R. 2004. *Islands of the Mind: How the Human Imagination Created the Atlantic World*. New York: Palgrave Macmillan.

Glick Schiller, Nina, Linda Basch, and Cristina Blanc-Szanton. 1992. "Transnationalism. A New Analytical Framework for Understanding Migration." In *Towards a Transnational Perspective on Migration: Race, Class, Ethnicity and Nationalism Reconsidered*, edited by Nina Glick Schiller, Linda Basch and Cristina Blanc-Szanton, 1–24. New York: New York Academy of Sciences.

Hay, Peter. 2006. "A Phenomenology of Islands." *Island Studies Journal* 7, no. 1: 19–42.

Ivin, Lovro. 2009. *Olib – Otok, selo i ljudi*. Zagreb and Olib: Družba Braća hrvatskoga zmaja, Mjesni odbor Oliba.

Jurić, Ljubomir. 1910. *Silba*. Zadar: Brz. katol. hrvat. tiskare.

King, Russell. 2008. "Geography, Islands and Migration in an Era of Global Mobility." *Island Studies Journal* 4, no. 1: 53–84.

Lajić, Ivan. 1992. *Stanovništvo dalmatinskih otoka: povijesne i suvremene značajke depopulacije*. Zagreb: Consilium.

Levitt, Peggy, and Nina Glick Schiller. 2004. "Conceptualizing Simultaneity: A Transnational Social Field Perspective on Society." *International Migration Review* 38, no. 3: 1,002–39.

Magaš, Damir, and Jadranka Brkić-Vejmelka. 2013. "Prirodno-geografska obilježja Silbe." In *Otok Silba, prirodno i kulturno blago*, edited by Jasmina Mužinić and Jenő J. Purger, 19–38. Zadar: Sveučilište u Zadru.

Mendas, Zrinka. 2015. "Tracing Socioeconomic Impact of Ferry Provision in Zadar Island Archipelago." *Journal of Marine and Island Cultures* 4: 10–26.

Mirošević, Lena, and Josip Faričić. 2013. "Demografska obilježja otoka Silbe." In *Otok Silba, prirodno i kulturno blago*, edited by Jasmina Mužinić and Jenő J. Purger, 251–67. Zadar: Sveučilište u Zadru.

Nikačević, Galeb. 2018. "Sućuraj na Hvaru: mesto gde letuje alternativna muzička scena bivše Jugoslavije." *Vice*, 1 September 2018, https://www.vice.com/rs/article/d3emdz/sucuraj-na-hvaru-mesto-gde-letuje-alternativna-muzicka-scena-bivse-jugoslavije (accessed in January 2019).

Nikočević, Lidija. 1983. "Rad mladih istraživača na Silbi." *Etnološka tribina*, 11–12, no. 4–5: 7–9.

Oroz, Tomislav, and Urem, Sandra. 2015. "Od 'otoka koji tone' do 'američkog arhipelaga.' Fragmentarna etnografija povratnih migracija na Olib." *Etnološka tribina* 45, no. 38: 122–38.

Petrinović, Kruno. 2006. *Kad Miki kaže da se boji: prilozi za biografiju Johnnyja B.* Štulića. Zagreb: Umjetnička organizacija Uradi nešto.

Rudan, P., A. Sulojdžić, A. Chaventré, and N. Borot. 1987. "Présentation de l'étude anthropobiologique de l'île de Silba: Aperçu géographicue, historique, ethnologique et démographique." *Rad Jugoslavenske akademije znanosti i umjetnosti. Razred za prirodne znanosti* 22: 237–53.

[s.n.]. 1953. "Uspješno poslovanje ovčarsko-mljekarske zadruge Olib." *Mljekarstvo* 3, no. 9: 214–15.

Srhoj, Vinko. 2014. *Šime Vulas: Jedreni totemi*. Zadar: Narodni muzej Zadar.

Starešina, Petar. 1973. "Kolonat na Silbi (u povodu 120. obljetnice otkupa otoka)." *Radovi instituta Jugoslavenske akademije znanosti i umjetnosti u Zadru* 20.

Starešina, Petar. 1971. *Pomorstvo Silbe*. Zadar: Jugoslavenska akademija znanosti i umjetnosti u Zadru.

Starešina, Petar. 1963. "Kapetan Bartul Supičić Silbljanin." *Acta Instituti Academiae jugoslavicae et artium in Zadar* 10: 417–26.

Tomljenović, Mladen. 1983. "Ribolov na Silbi." *Etnološka tribina* 11–12, no. 4–5: 45–48.

Vigato, Ivica. 2013. "Povijest Silbe: bilješke iz ostavštine Petra Starešine." In *Otok Silba, prirodno i kulturno blago*, edited by Jasmina Mužinić and Jenő J. Purger, 165–74. Zadar: Sveučilište u Zadru.

Chapter Eight

Two Islands in One

The Island of Hvar

Ana Perinić Lewis and Pavao Rudan

FIRST ISLANDERS AND NEWCOMERS

Island space is special in comparison with the mainland. It presents a double challenge because of its specific geographic characteristics and the notion of territoriality, which has been contested throughout history. Island and mainland spatial perspectives are different. Islanders rarely define their island as part of the mainland or continent, and they never identify it with neighbouring islands. Apart from the fact that they are all surrounded by water, each island is unique (Skračić 1997). The concept of boundaries on an island is extremely interesting. In researching islands, most attention is usually paid to their external boundaries, coastlines or rims as the essential factors that constitute islandness. Islanders tend to be more in contact with and aware of these boundaries than most other people (Hay 2006).

While the external boundaries may be fixed, internal boundaries are often subject to change, adoption and the remodelling of the island space as a whole. Divided islands are usually understood to be those whose territory is crossed by an international border or those shared between two or more countries. The most obvious European examples of shared island spaces are Ireland (see Chapter 1) and Cyprus, which belong to a group of 10 islands under different jurisdictions (the others are Usedom/Uznam, New Guinea, Borneo, Timor, Hispaniola, Sint Maarten/St Martin, Isla Grande de Tierra de Fuego and Bolshoi Ussuriysky/Heixiazi) (Baldacchino 2013). Although the islands in the Croatian archipelago have belonged to different states during their history, there are no divided islands among them today, but there are examples of separated or fragmented islands. The island of Pag, as mentioned in previous chapters, is a special case.

Proximity to the mainland and its influence on coastal islands and their islandness is different and unique for every island in the Croatian archipelago. Some islands and subarchipelagos have failed to develop strong interisland links and firm identities, and often, proximity has not been beneficial, but has sharpened interisland division and led to increased isolation. Although newcomers on the islands are a relative concept, since they have always existed (Duboković Nadalini 2001), each island has its own indigenous inhabitants, the people who first discovered it, settled there and acquired or created the status of "first islanders" or natives. Anyone who came after them, in groups or individually, usually settled in parts of the island set apart for immigrants or established completely new island settlements.

An awareness of newcomers, those who arrived later, constantly emphasising and highlighting their otherness, is also part of island history, heritage, culture and identity. "First islanders" never forget where the newcomers came from (which part of the mainland or the sea route they took), nor when they arrived, and can never regard them as "real islanders." "Other" islanders and island community are perceived as having a diverse, often detailed spectrum of distinctive features, real or imaginary differences, boundaries and frameworks that define, separate or isolate them. In this chapter, we will explore the internal boundaries that characterise one particular island and its settlements, and which affect the identification of islanders and various representations of islandness.

Boundaries, as places of demarcation and articulation, are not only physical, but also imaginary. According to Edward Said (1999), they form a constitutive part of mental maps, or imaginary geography, in which demarcation is based on separating what is ours from what belongs to others, and finding our place in the world. The boundaries between island settlements run over land and through the sea, and are extremely stable, enduring, complex and much more difficult to cross than external ones. Such boundaries divide the island population, and differences are seen in language, customs and mentality. Sometimes, the difference is obvious, but at other times it is only perceived by the islanders themselves. It is best illustrated by a claim made by Joško Božanić (1997): If an islander has to leave his village, he would rather live on the other side of the world than on the other side of his island.

We will also consider the identification of islanders linked with certain parts of an island or island settlements and stereotypes about these areas and their residents. Since the islands in the Croatian archipelago are close to the mainland and have links with it, it is important to realise how the islanders regard the mainland and its population. In the eyes of most mainland residents, all islands are the same. In the same way, islanders have a homogeneous view of the mainland, seeing it as one area and dismissing its variety. They see mainlanders as "others" who are divided from them by the sea. The oldest contrast between island and mainland populations (which also occurs in intraisland divisions) is preserved in the names *Vlah/Vlaj* and *Bodul*. The division of the population into those living by the sea and those in elevated areas is characteristic of the entire Mediterranean (Braudel 1997; Matvejević 2007).

In the eastern Adriatic, the oldest, most enduring stereotype, of which we find confirmation in historical sources from the 14th to the 18th century, dubs the inhabitants of the mountainous hinterland as Vlachs or Morlachs (Morlach—Greek *Maurovlahos*, Latin *Morlachus*, Italian *Morlacco*, or Black Vlach, the name for the population of Venetian Dalmatia in the hinterland of the eastern Adriatic coastline.) Today, the name Morlach is an archaism; however, the stereotypical name *Vlah* in Dalmatia and the Dalmatian islands has been retained as a geographical and social determinant (Botica 2005; Kovačec 2016). While this name is used on the mainland for those in the interior of the continent, on the islands it is mostly used for nonislanders and any continental dwellers, whether they live in the coastal area or the mountainous hinterland. Islanders also use it for those who moved to the islands during the last centuries (Kovačec 2016). The name is also used for those who do not understand the sea or ships (Skok 1973; Barbić 2012; Benčić 2013). Thus, the Croatian islanders emphasise their exclusivity and distinctness, and think of themselves as the only true Mediterranean people, of the Adriatic Sea specifically, while casting doubt on the "Mediterraneanism" of the mainland population.

On the island of Hvar, *Vlah* includes all these meanings, both within the island and in the island–mainland relationship, as presented here and confirmed by field research. The opposite is *Bodul*, which is what mainlanders call islanders. The name first occurred in the Kvarner area, where the islanders accepted it with pride. The inhabitants of the central Dalmatian islands do not use or accept it, as they think it is used pejoratively by mainlanders. On Hvar, it is obvious that in their identification strategies the islanders are not content with merely using *Vlah* to describe a continental dweller but also apply it to those living in the hills and uplands on their own island. Island highlanders (*gorštaci*), as referred to by some islanders, are proof that islands create their own binary opposites and form separate but by no means homogeneous units. The basic Mediterranean dissimilarity between the sea and the mountains, between the people from the coast and highlanders, is apparently sufficient to form a microregion on a very small area of land surrounded by the sea.

Each inhabited island in the Croatian archipelago has some more densely populated and developed areas, with historical continuity in the formation and development of settlements and the population. Due to geographical, ecological and historical factors, some island areas are less attractive, more isolated, less accessible and sparsely inhabited. Throughout the centuries, they have often been inhabited by displaced people, newcomers or immigrants who were different from earlier settlers. Along with the tenant-farmer/landlord system and migration, diachronic changes in the administrative-territorial divisions of the Croatian islands between countries, regional and local centres have affected the division of island spaces and the creation of distinct communities and settlements. This has led to fragmentation and a new type of insularisation within the island space, settlements and population. As a result, communications and contact between islanders, and neighbouring islands and the mainland, have often been disrupted or hampered.

THE ISLAND OF HVAR

We chose the central Dalmatian island of Hvar to study real and imaginary island boundaries. It is the longest Croatian island (68 km). In contrast to the Dinaric, which lie from NW to SE, Hvar is specific because it extends from west to east and narrows toward the western tip. Strange as it may seem to "continentals," its shape actually affects the spatial relationships of the Hvar population groups, particularly their use of spatial adverbs to describe moving around the island or determine the positions of island regions and settlements.

Hvar is characterised by a predominantly hilly relief composed of three main units: the ridge or "spine" of the island, which forms most of the island area; the fields or the "belly" (comprising Starigrad and Jelsa Plains, the sites of the earliest organised settlements) and the Rudine-Kabal hilly tract in the northwest part of the island (Bognar 1990). Today, connectivity between Hvar and the other islands in the central Dalmatian archipelago is weak. It is retained only in the three-island Diocese of Hvar, Brač and Vis, with the diocesan see in Hvar. There are no strong links with neighbouring Brač, apart from ferry lines established in recent years. This is in stark contrast to the strong historical links between Hvar and Vis. These connected Pharos (Hvar) and Issa (Vis), the two oldest Greek island colonies on the eastern Adriatic coast, along with the Vis-Hvar alliance between Illyrian queen Teuta and Demetrius of Hvar, who challenged the Roman Empire. Later, Vis became part of the Hvar commune under the proprietorship of the Hvar landlords, which lasted several centuries.

We analysed identifications and representations of local identities on the island of Hvar, the division of the island and island space, and its inhabitants. The morphology forms a natural boundary between the eastern and western parts. Throughout history, natural factors have strongly influenced the location and types of settlements, and limited communication between villages. The eastern part of the island is a rocky area with scattered settlements, while the western part has been continuously densely populated.

The island is thus divided into eastern and western microregions, and this has led to the division of the population between the western, indigenous, Mediterranean inhabitants and the eastern immigrants. The eastern karst is shaped like a long blade that narrows toward Sućuraj and accounts for 54% of the island's length and about 42% of its surface area. The karst is exposed to heat and drought in summer, and cold onslaughts from the mainland in winter (Roglić 1977). Its inhabitants are stereotypically described as island "highlanders," "non-Mediterranean" and even "nonislanders." When did the Hvar islanders begin to distinguish the more prestigious parts of the island from the less attractive areas, reserved for undesirables and nonislanders? When did they begin to construct complex systems for dividing the island space and people, apart from its external, sea boundaries, which have lasted until today? The turbulent history of the island is evident in its historical names: Pharos, Pharia, Lesina, Hvar and the names used by locals, Far and For.

The island's prehistoric inhabitants were bearers of Croatian Neolithic culture (6th to 4th millennium BC), which was dispersed throughout the entire island and has been preserved in 17 archaeological sites (13 of which are caves) (Gaffney, Kirigin, Petrić, Vujnović and Čače 1997; Forenbaher 2002). According to archaeological finds indicating their life and culture in the two best-researched locations (Grapčeva Cave on the south side of the island, near the village of Humac, and Markova Cave on the northwest side, on Pelegrin Point), we know that they engaged in livestock farming, agriculture, hunting and fishing (Novak 1955; Zaninović 1995). They also made delicately painted pottery (Petrić 2002; Forenbaher and Kaiser 2008) and were actively linked to ritual activities concerning the dead (Forenbaher, Kaiser and Frame, 2010). The indigenous inhabitants were known to the ancient Greeks as Illyrians. Their settlements were often located on elevated, easily defended positions. The remains of about 15 hillforts (*gradine*) or fortified settlements can still be seen on Hvar, spread throughout the entire island from the town of Hvar to Sućuraj (Gaffney et al. 1997).

Immigrants who sought and selected the "best" Adriatic islands and then chose particular island locations among them were the most influential in terms of long-lasting, organised settlement. They came from the Greek islands of antiquity, Syracuse and Paros. They founded the first Greek colony, Issa, on the island of Vis in 397 BC, while in 385 BC, colonists from the Aegean island of Paros founded Pharos (on the site of modern-day Stari Grad) on the island of Hvar (Novak 1960). Recessed Stari Grad Bay was suitable for building a harbour and also situated on Adriatic trade routes, next to the largest fertile plain on the island. The *Chora Pharaou* (Stari Grad Plain) covered approximately 1,200 hectares, and 73 of its regularly arranged parcels, with public paths leading to the town, are still preserved today (Kirigin and Slapšak 2010; Solarić and Solarić 2009). The people of Pharos fortified their newly founded town with ramparts and initially allowed "barbarians" to live in their inaccessible fortress (Diodorus of Sicily XV, 14, 1–2, according to Čače 1997).

The first distinction between "us" (Greeks) and "them" ("barbarians," "Illyrians") with whom they had conflicts in their newly founded colony was reflected in the division of the island space between the prestigious, attractive areas (usually on the shores) and the less attractive, hilly areas where "others" were located. After defeating the Illyrians, the people of Pharos drove them to the eastern side of the island. Thus, eastern Hvar became a place to which undesirables were exiled; however, the Illyrians found this part of the island quite conducive to living and waging war, as they could control the coast and sea. This is evident from the many burial mounds they left behind (Dragičević 1970; Radovanović 1976; Gaffney et al. 1997).

Illyrian commander Demetrius of Hvar was the first famous inhabitant of the colony of Pharos to be mentioned in historical sources. Through his military and matrimonial alliance with Illyrian queen Teuta, he ruled the Illyrian state and fought the Romans. In 219, the Romans conquered the island. Roman Pharia did not become a municipality but an area within the province of Dalmatia administered from Salona, a mainland town (Novak 1960). For the first time, the island came under direct administration from the mainland. In the 8th century, a new immigrant group, people from Neretva and Croats, Slavicised the island, as can be seen from toponyms. In 1278, Hvar came under the rule of the Venetian Republic, which called it Lesina. The central Venetian administration of the island and commune was located in the town of Hvar, which was fortified and became a winter base for the Venetian navy. Hvar Town developed as the main island centre, taking over the former role of Stari Grad.

The Statute of the medieval Hvar commune (1331) regulated legal relations on the island and was also important because it recorded the number of island settlements, including the eastern part of the island, for the first time. It also regulated the oldest administrative island boundary and the historical-administrative name of Plame (*confines de Plame*) for that part of the island. The demarcation line ran southeast from Jelsa (Mina Bay) to Sućuraj (Sveti Juraj) on the eastern island cape (Duboković Nadalini 2001). The actual statutory boundaries of Plame (particularly the boundary running inland) were rather ill-defined in terms of medieval locations, their names and existing settlements at the time, and in relation to contemporary settlements. In addition, Plame's size and boundaries varied throughout the centuries. According to the Statute, it was municipal property, an area where no one was allowed to acquire personal property, reserved for pastureland, logging and growing cereal crops (Kovačić 1998). There were no permanent settlements, and court cases mentioned only those who were there for temporary work; however, since the central part of the island became overpopulated and because of the arrival of newcomers from the mainland, settlement was encouraged, and in the 1407 revision of the municipal land register, the first private proprietors in Plame were recorded (Kovačić 1995).

Beginning in the 15th to the 18th century, the Venetian Republic ruled Hvar. This was during the era of the Turkish wars, which prompted mass population migrations, and Hvar became a refugee destination and safety zone. Plame was settled by newcomers from existing settlements in the west of the island (Pitve and Vrbanj) and, after the Turkish wars, by refugees from the mainland (*abitantni nuovi*), who created settlements under the administration of the Venetian Republic and its political-legal system, which included various privileges for newcomers and indigenous islanders.

In the 15th century, the villages of Zastražišće and Poljica sprang up as permanent settlements, along with Gdinj and Bogomolje. In the 17th century, the Venetians encouraged the Christian population from Turkish areas to emigrate to mainland Primorje and the Dalmatian islands, with the aim of creating a detachment of border guards and ensuring military protection (Čolak 1959; Lajić 1992). The Venetian Republic conducted planned colonisation by nonislanders, attracting them with economic and political privileges, *Privilegije Paštrovića* ("Paštrović Privileges"), which they enjoyed until the arrival of the French authorities in 1804. They acquired land to cultivate and were exempted from paying customs and tributes, serving on Venetian galleys in wartime, paying certain taxes and performing public works. These privileges were based on the concession model the Venetian Republic granted to the Paštrović clans from the Montenegrin coast in 1423, when the area came under Venetian control.

Privileges were individual, so that each newcomer had a document proving his rights. Privileges were withdrawn if new residents refused military service or married the indigenous people. This was a cause of division and conflicts between groups of old and new residents (Čolak 1959; Duboković Nadalini 2001). Privileges were defended and guarded, as is clear from many archive records of meetings, elections of representatives and issues contested before the authorities of the Hvar commune (Čolak 1962; Novak 1961; Bezić-Božanić 1988). The *abitanti nuovi*, as they were called by the Venetian administration, or *Primorci*, as they appear in archive records, came to the island in large groups in the 17th century and settled in almost all its settlements.

There were few of them in the eastern part of the island, where there were either no indigenous people (Sućuraj, a new village settled by newcomers) or only a few (Gdinj and Bogomolje). They became the dominant population there (Duboković Nadalini 2001). They used their privileges to further distinguish themselves from the indigenous population to ensure that the eastern part of the island became more self-contained, different and distant from the urban, lowland, indigenous part. Plame was obviously always reserved for shepherds and newcomers, and all different "other" inhabitants of Hvar.

Thus, the island of Hvar became a paradigmatic example at the micro level of the chasm between continental newcomers and indigenous islanders, a place where the Dinaric and Mediterranean spheres met, where mainland and island Dalmatia were linked and where islanders and highlanders all lived. A range of stereotypes formed regarding "others" and the "other" part of the island, traced diachronically for more than 400 years.

FROM ROUGH "OTHERS" AND STRANGE, WILD PEOPLE TO "OUR OWN" ISLAND HIGHLANDERS

When did these new islanders become a presence in the island landscape and consciousness of the islanders? Stereotypes about those living in the east of the island appeared early on. In a speech entitled *De origine succesibusque Slavorum*, delivered by Hvar Dominican and humanist Vinko Pribojević in Latin in the town of Hvar in 1525 (printed in Venice in 1532), he described his native island from the east (nearest the mainland) to the west. He divided Hvar into three geographical areas with their inhabitants and economic branches: 1) the part of the island facing the mainland, with five settlements on level ground (shepherds); 2) the plain, which extended north through the central and western parts of the island, surrounded by 11 villages (labourers and fishermen); and 3) the south coast with the town of Hvar (the nobility and citizens). The inhabitants of the eastern part were described as shepherds who were "rougher" than the other islanders (Pribojević 1951, 199), contrasting them with the other people of Hvar and clearly identifying them as "others." The Hvar commune, with all its classes, was described as a harmonious, united, ideal community. The shepherds and their five villages on the highlands were excluded from collective representation.

Pribojević identified several groups (the nobility, the common people, the labourers and fishermen) and identities (Slav, Dalmatian and Hvar). He used positive stereotypes to affirm established relations and values, and presented the people of Hvar using the historical, linguistic and cultural branches of knowledge available to him. He characterised the inhabitants of the eastern part of the island as different from the "others," without giving a single indication of their differences, while insisting on their unambiguous roughness. The partially defined notion of who the Vlachs were (and in the eyes of the islanders, this meant anyone from the continent) was now affirmed realistically and in more detail on the island. The people of Hvar saw their own island highlanders in the immigrants from the mainland. In Pribojević's time, these new islanders came in the first wave of mass migration by refugees who found safety and a new home on the island. Settled in the karst, an area with an inclement climate, limited means of existence and weak links with the more developed Hvar commune and its culture and heritage, they were simply unknown foreigners, strange, alien and possibly unfriendly continentals. In the golden age of the island, the people of Hvar kept their distance, did not give them a name and certainly saw no reason for them to be integrated into or assimilated in island life (Perinić 2006).

Hvarkinja ili Komedija od Bogdana (The Woman of Hvar, or Bogdan's Comedy) was a Renaissance comedy by Hvar playwright Martin Benetović (1550–1607). It featured local 16th-century stereotypes connected with particular island communities. It was performed during carnival time on the main square in the town of Hvar (*Pjaca*), which was the stage and place of the action. The plot and characters were a reflection of social and class relations, and life in Renaissance Hvar (Bogišić 1977). Since this kind of comedy created types rather than individualised characters, stereotypes played an important part in characterisation.

Through the relationships between the characters in the comedy, Benetović depicted the island communities as microcosms with clearly marked boundaries. Relationships between peasants and citizens were a source of humour in Renaissance comedies and are now a source of information about the social and economic status of the peasant class in society. Benetović introduced a name for one inhabitant of the eastern part of the island that is still present in Hvar speech today—*Bogdan Plamjanin*. The peasants in the play can be clearly divided into two groups: *Plamjani* (peasants from Plame) and *Vlasi* (peasants from the mainland, from the mainland Makarska Primorje area). There is a clear hierarchy. The *Plamjani* are the "other" newcomers who have settled the island karst area; however much they have kept their highland customs, language and traditions, they still live on the island. On the other hand, the *Vlasi* are peasants from the mainland who are completely unaccustomed to not only the urban environment, but also island life itself. The peasants and servants from Plame, comic characters destined to be laughed at and mocked, cease to be objects of mirth when they encounter the *Vlasi* and join the other island classes in mocking and stereotyping the mainland peasants. Although the *Plamjani* themselves originally came from the mainland, they are less foreign to the people of Hvar than the *Vlasi*. In the encounter between the two groups, the *Plamjani* become equal members of the island collective, adopting the Hvar collective *imaginarium* and the islanders' derision of continentals. A *Vlasi* stereotypical view of the inhabitants of the island and coastal towns is also introduced in the play. Radoj and Barbara, who are originally from the mainland, call the people of Hvar "Latins," because they think their customs and culture are foreign. In Benetović's comedy, the rough, nameless shepherds of Pribojević's Hvar become *Plamjani* and are among the main characters, representing a particular category of resourceful island highlanders.

Quite early on, a reverse stereotype also appeared—the "view from the east of the island." It is found in a late 16th-century Hvar comedy by an unknown author, who literary theorists assume was an imitator of Benetović. *Komedija od Raskota* (The Raskot Comedy) is about people from Plame, for people from Plame, set in Plame. Plame is referred to as "our mountains" and "our highlands," and local names and places are mentioned. It is as though the western part of the island, with its urban life and various classes, does not exist. In this comedy, the scene of the action has no characteristics of an island or Mediterranean place, and the sea and island are rarely mentioned. It was written in response to the need of the Hvar public for plays that could be performed in Lent, located outside the urban world of the Hvar commune, and the anonymous author wanted to satisfy the audience's interest in people from Plame (Perinić 2009).

In the latter part of the 18th century, an Enlightenment circle of friends was formed on Hvar. It was founded by three Italian and Dalmatian adherents of the Enlightenment who had included their experiences of the island area and individual Hvar communities in their creative opuses: Alberto Fortis (1741–1803), Julije Bajamonti (1744–1800) and Ivan Dominik Stratiko (1732–1799). All three spent time on the island and had firsthand knowledge of Hvar and its people. They were well educated and had close links with Dalmatia, either by birth (Bajamonti was born in Split and Stratiko in Zadar) or in a spiritual sense (Fortis was born in Padua in Italy). They described and explained their views of individual Hvar settlements and their inhabitants in the 18th century.

Fortis was the famous Italian Enlightenment travel writer. He visited Hvar (9–20 December 1772) during his travels in Dalmatia and recorded his description and experience of the island in *Viaggio in Dalmazia* (Travelling Across Dalmatia). He mentioned famous people from Hvar in his description of the island and its history. He described the island settlements (Hvar, Stari Grad, Vrboska, Jelsa and Sućuraj) and inhabitants. He criticised the citizens of Hvar Town for being friendly toward foreigners but not to one another, while he found the aristocrats and citizens of Stari Grad to be men of stature, courageous, clever navigators and ship-owners, and the common folk shipbuilders or fishermen. He highlighted Jelsa and Vrboska as welcoming fishing villages. Sućuraj was the only eastern place he mentioned, and he found it unprepossessing and underpopulated. He had nothing to say about its inhabitants (Fortis 1984). The person who provided most of his information was his close Dalmatian friend, Julije Bajamonti, who was the doctor on Hvar for five years. Ivan Dominik Stratiko spent the last years of his life on the island as the bishop of Hvar (1784–1799).

Bajamonti and Stratiko wrote two letters in Italian about the island of Hvar and its inhabitants. The letters are unique in portraying the island space and present two imaginary worlds inhabited by indigenous islanders, one of which describes the "Mediterranean" population and the other the Hvar "highlanders." Bajamonti accompanied Fortis on his travels, supplied him with information and was a close friend, as can be seen from their surviving correspondence (Milčetić 1912). He wrote *Lettera sopra alcune particolarità dell' isola di Lesina* and sent it to Fortis on 25 May 1790, with the request that it be printed secretly at his own expense. It consists of several thematic paragraphs. After a short description of the position and climate of the island, Bajamonti writes from a doctor's perspective about the health and sicknesses of the people and their mortality rate. According to his observations, they enjoy exceptionally good health, so he has little work. He provides an eyewitness account of what it was like to be a doctor in 18th-century Hvar, describing his routine and the unpleasant things that were part of his life as a doctor in a small community. He is scathing about the number of certificates he was required to issue: Clearly, the citizens of Hvar were prone to litigation and violence.

> In less than five years, I have issued about a hundred doctor's certificates and criminal reports (on violence or violent deaths), and have avoided as many more which I thought unimportant. Among these people, the issues are not always murder, poisoning or other serious cases: They are usually about blows, scratches and similar, particularly between fathers and sons, or husbands and wives, not counting jealous outbursts, which are usually resolved without involving the doctor or the courts. (Quotations taken from a translated transcript of the entire letter in Italian published in Novak 1978.)

The sharp-eyed doctor was also surprised by a large number of foundlings in the town and had some ideas about how they came to be there. Simply by choosing to write about foundlings in his description of the population, Bajamonti hinted rather slyly that they were not overly moral people. He also writes about the medieval Hvar Statute, from which he selects the most interesting or most bizarre extracts. He also comments on contemporary Hvar and its inhabitants.

> Today, the Hvar nobility read, write and behave like gentlemen, not because the law requires them to, but because they are personally inclined and educated so to do. Today, one can live in Hvar without keys or bolts. There is only petty theft of fish, grapes (muscat) and other sweet fruit, which are often stolen while still unripe.

There follows a comparison of the "golden age" of Hvar, as described in the speech by Vinko Pribojević (1525), and its decaying opulence and significance in Bajamonti's time.

> Now, the harbour is almost always deserted: so farewell Scythia, farewell Arabia, farewell eloquent Greece, farewell all other lovely things linked with the departed past. However, a 16th-century Hvar friar does not mention that there were once on this island things that came from refined Gaul, and 18th-century Hvar doctors can curse that sad fact today.

The paragraph ends with a barbed comment about syphilis, the "French disease," pointing to it as a "popular, French export" found on Hvar. The final paragraph deals with ecclesiastical history taken from historical and church sources, with anecdotes about the Hvar bishops and priests.

As a true patrician, he retained his Enlightenment optimism and made friends in spite of the blinkered bourgeoisie and prejudices he encountered. He ends the letter by praising friendships made in the Hvar environment.

The third member of this Hvar Enlightenment circle was Ivan Dominik Stratiko, bishop of Hvar from 1784 to 1799. When he arrived, he was already a world-renowned Enlightenment scholar and one of the leading intellectuals of his day. He was born in Zadar and became an erudite member of Italian academies (Dei Rozzi in Siena, and Degli Arcadi in Rome), and a friend of Italian and Dalmatian Enlightenment adherents. He knew Fortis's work well, and "Morlacchism" (a literary trend in Europe at that time based on the European reception and exoticisation of the hinterland Dalmatia and their inhabitants, Morlachs, and their customs, beliefs and way of life). During his first episcopal visitation on 1 April 1786, he made many careful notes and kept a copy of a private letter sent to his friend, Ansano Luti, in Siena. It was dated 24 May 1786, and described Stratiko's visit to Bogomolje on eastern Hvar (Krasić 1991). The Hvar bishop discovered there his own, private Morlachs in his own island Morlachia. He uses Morlach discourse in the letter, faithfully following the Morlach trend, and applying it to the people of Bogomolje. Writing to his friend, who was well acquainted with Fortis's *Viaggio in Dalmazia*, like most intellectuals of the time, he described the meeting between the new bishop and his parishioners in line with the established assumptions and expectations of the Italian public regarding exotica from the eastern Adriatic coast. He describes some of his flock using Enlightenment concepts of the "noble savage." The people of Bogomolje are simple, virtuous and good, but their barbaric, savage nature is emphasised in the very first sentence: "I do not know why writers always give examples of barbarians from Canada or Madagascar, when these very near islands, at a much closer distance, provide so many." (Quotations taken from the original letter by Ivan Dominik Stratiko to Ansano Luti of Siena, 24 May 1786, Biblioteca Comunale degli Intonnati Siena CSI D. VII. 21, f. 254. The original letter was published in Perinić Lewis and Rudan 2011, 86).

Stratiko (1786) and Fortis (1774) both stress the people's poverty and backwardness ("They have little reason and use it less"), instincts, lack of moderation, lack of economy and the position of women in the community ("The women are treated like dirty animals which are not mentioned unless someone says, 'My wife,' apologetically. They do not dare to be where the men are, unless invited"). Like Fortis, Stratiko reports to his friends in Italy as though he is travelling in a new, undiscovered land. The exception is his description of the attitude of the Bogomolje people to the clergy. He seems to adopt the position of an "insider," as his description of the experiences of the clergy in the community of the faithful slips into the first person singular.

> Wherever I go, they carry me on an open litter like a sort of saint, and I am surrounded by guns and sheathed sabres. The men salute me in a soldierly way at all stations, and the women follow, singing and dancing. They consider themselves blessed if they can touch my garments, kiss my ring or receive a rosary, medal or relic from my hands. The most violent arguments cease at my bidding, what is hard to give back is given back, and the most soiled affairs are settled.

Fortis noted and pointed out the differences between the populations of coastal towns and the hinterland throughout Dalmatia. Stratiko noted a similar divide between the inhabitants of western Hvar with its urban centres and those living in the eastern settlements.

> Since there are many men living in the town who have sailed the seas, they bring back with them certain Italian customs and habits, but they are in the minority. The people of small towns, villages and the mainland care little for the town, and rarely go there, they have nothing in common, neither language, nor religion, nor costumes, nor the laws of civil society.

An important feature of Stratiko and Bajamonti's discourse is their ambivalence, whether writing for the educated European elite to which they themselves belonged, speaking in front of Dalmatian intellectuals or writing motivational articles to improve circumstances in Dalmatia. In their works, they put Dalmatia on the map of the European Enlightenment. Fortis "discovered" it, and Bajamonti and Stratiko affirmed and broadened his *imaginarium*. On one hand, they complemented the Venetian imperial discourse on Dalmatia and helped construct another (of the Morlachs, Schiavoni [Slavs], Dalmatians and Croats) in relation to the Italians in the age of the Enlightenment. On the other, both actively participated in the physiocratic movement and reforms in Dalmatia that aimed to improve the economy and the education of the population. The fact that the island centres were not prepared for people with big ideas was demonstrated in letters from the priest on Korčula, reporting Bishop Stratiko to the Vatican in 1792, for belonging to a masonic brotherhood (Perinić Lewis and Rudan 2012). Bajamonti's letter about Hvar and its people was not forgotten in the 20th century. Well-known Croatian historian Grga Novak mentioned that he had not included Bajamonti in his history of Hvar because the latter had not understood the Hvar milieu marked by tradition and family links (Novak 1978).

Examples of this discursive formation of island stereotypes about "otherness" proved enduring (from the 16th to the 18th century) but also variable. Stereotypes about the Hvar highlanders became more sophisticated and precise throughout time, and more frequent, involving new relations, creating new, clichéd characters and prescribing new contexts of use. In spite of their variability, their permanence has become an integral part of the Hvar island identity.

BORDER GAMES: WHERE EXACTLY DO THE "OTHERS" BEGIN?

Having looked at the sources and literary works where the "others" appear as the eastern inhabitants of Hvar, we decided to carry out some field research to see if the stereotype still persists today, and if so, in what form, and with what changes. The aim was to explore the complex identification processes of Hvar in the 21st century, based on the division between the eastern and western microregions, the assumptions both groups of inhabitants make about each other, the boundaries they set and the effect of these perceptions and boundaries on how they interact with one another. We analysed the ways in which they presented their own group and the "others," and how they construed the identity of the neighbouring group as "other" using attributes relating to differences and stereotypes. We carried out comprehensive field research on Hvar between 2006 and 2008, among 100 respondents of different ages from all the permanently inhabited island settlements. Sixty respondents were from the west part of the island and forty from the east part. The identity of the respondents has been protected by using aliases, but the details regarding their places of birth and residence, and their ages, are correct, unless they themselves requested they should not be mentioned.

It was interesting to discover how enduring the boundary separating the parts of the island really was. Plame, named and defined in the Statute of the medieval Hvar commune, does not exist today in the administrative-legal sense, but it is still very much alive in the island *imaginarium* and mapping of "otherness." The inhabitants of western Hvar continue to use the historical name Plame for the eastern part of the island, usually placing the boundary in the same place as the one defined in the Hvar Statute. Some even appeal directly to the Statute as the valid document that affirms the boundary.

> Frane (Hvar Town, 50): "Well, everybody knows, the borders of Plame were clearly marked in the Statute. So, from Poljica on."

The most frequent answer to the questions about the settlements and people living in the eastern part of Hvar was to place them geographically using such phrases as "Everywhere above/below Jelsa," "Higher up than Jelsa," "Up the hill" and "Above Jelsa," or even "At the island's tail" (see Figure 8.1). The people in the east part responded reversely. The answers about the west part were "Down," "Down there" and "Down toward." Twenty-three respondents used the adverb "far" or "distant" to describe where the other part of the island was, but only four (three from the west and one from the east part of the island) were able to tell the exact distance.

However, the inhabitants of some villages drew different borders for the eastern part of the island, particularly those from the villages around Starigrad Plain and Jelsa. They did not include Poljica (and, more rarely, Zastražišće) in Plame and the stereotypical names used for its inhabitants. Most of the inhabitants of Jelsa or villages that gravitate toward it strongly did not say that Plame began "beyond Jelsa," because that would mean they themselves lived on the border. Instead, they moved it to "from Poljica on" or "from Poljica and Zastražišće." So, the people of Plame (*Plamusi*) would be from the three settlements on the eastern side of the island: Gdinj, Bogomolje and Sućuraj. The inhabitants of Hvar Town set a separate boundary, because to them, Plame is everywhere outside their town. Some include Jelsa (see Figure 8.1). The most extreme boundary is set by some individuals from the town of Hvar, who think Plame begins right where the old town core ends (beyond the walls of medieval Hvar).

The boundary is usually moved just far enough to make sure that one's immediate neighbours are not *Plamusi*. Today, there is still a vivid notion of the separate, eastern part of the island, its inhabitants, villages and way of life. The most extreme form is a stereotypical identification of the people of Plame with the Vlachs, a name often used on the islands for mainlanders. On Hvar, it means their own island Vlachs.

> Kuzma (Svirče, 52): "But we have so little in common, there are two different worlds, so I think hardly anyone would have much to say about that part of the island, apart from the fact that we consider them to be our own island Vlachs."

The people of Hvar Town do not even regard the newest island settlement, Sućuraj, and its inhabitants as "highlanders." They emphasise their strong links and similarities with the mainland from which they came, rather than with the island space. They point out that they use the Shtokavian dialect spoken in the Makarska coastal area (in contrast to the people of Hvar, who mostly use the Chakavian dialect), so they do not consider them real islanders.

> Pavica (Vrbanj, 43): "They are different people! No, they don't even consider themselves to be islanders, because they are oriented towards Makarska and Drvenik. So it's easier for them to go there than to Hvar Town."

Throughout time, many variations of the mocking collective nicknames for the people of the region have developed (*Plamjani*, *Plômjani*, *Plôvjani*, *Plami*, *Plamusi*, and *Plamusori*), but the most common one today is *Plamus*, used in the western settlements in such phrases as "You walk like a *Plamus*!" or "Walking like a *Plamus* going to court." The eastern inhabitants of Hvar came under the jurisdiction of the court in Stari Grad, which they could only reach after many hours walking, because of the bad roads and long distances. They used to walk in single file, and this may be the origin of the second phrase. Two interpretations are the most likely. The first refers to opposing parties in a court case who were obliged to walk together to Stari Grad. The other includes another derogatory stereotype about the easterners and implies that they were not used to walking on roads, but paths through the fields, so even when they went to town, they continued to walk in single file.

> Ivo (Stari Grad, 58): "Up there they only had paths, tracks, so narrow they had to walk one behind another, not side by side. And when they got to a wide road, they just kept on walking one behind another, because that's what they were used to. So we say, 'Walking like Plovjani!'"

The inhabitants of the eastern villages knew the nickname was derogatory and bestowed on them by others. Since they had no way of combating the stereotype, they accepted it in a way, explaining it as a fact of life they would have to reconcile themselves to.

> Danijel (Gdinj, 67): "In a way, we are the hill-dwellers on this island, as far as the people down there in Hvar, Stari Grad and Jelsa are concerned. So people a bit further off, further away."

These statements show a kind of acceptance of imposed nicknames and categories, even when derogatory. The experience of being belittled and degraded unites people. They are resigned to regional nicknames and stereotypes that have lasted centuries and become rooted in the island *imaginarium*. Such names were bestowed by the indigenous people on newcomers who even today realise that they cannot claim to derive their origins from Greek Pharos, but from refugees from the mainland. These examples reveal the uneven balance of power between the two communities, since the community whose members form the island centres of power and authority is the same one that has created stereotypes and belittled its neighbour.

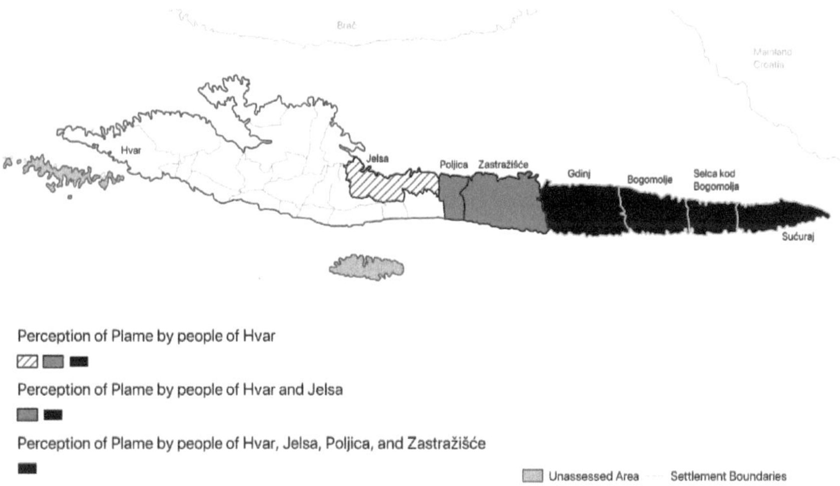

Figure 8.1. The Island of Hvar

The eastern island inhabitants also created mocking nicknames, stereotypes and proverbs for their western neighbours, although these have not been recorded in research to date. Although the westerners assigned them negative attributes and stereotypical names, and identified them as antipodeans, looking only in one direction from their centre of power, the easterners also resorted to subverted, imposed classifications, creating stereotypes about their "powerful" neighbours. The results of field research in the eastern Hvar villages shows that stereotyping works two ways, as does interpreting the differences on which to base an identification strategy.

The most frequent derogatory name is *Donski* (Lowlander). It derives from the opposing notions of "up" and "down," which also indicate east and west, the rising and setting of the Sun. It refers to all the inhabitants of the western part of the island, uniting their differences and specifics, and unrelated to any particular geographic reference point. It simply expresses the people who live "down there," "below Jelsa" or "out west." There are many explanations as to why one name fits them all. It is often said that they use different dialects that the easterners cannot understand. They are said to have strength and stamina but not too many brains, to stress the intelligence and quick thinking of the people from Plame who live in the western part of the island today. *Donski* are thought of as "distant," because they have always held authority and power, and usually exploited or abused the easterners.

> Margarita (Gdinj, 58): "The Donski folk have always been distant. Because they were on the lower part of the island, near the sea, in three or four places, so they were always differently connected with ships from Split, and we were cut off from them and from the mainland, and that's how we've stayed."

Two other pejorative names go beyond the framework of the island regions and villages: *Boduli* or *Talijani*. *Boduli* is the word used by mainland dwellers for islanders. The inhabitants of the Dalmatian islands, in contrast to the Kvarner islands, neither use nor accept this name and refer to themselves simply as islanders. So the people of eastern Hvar use the pejorative mainland nickname for the western islanders. They also call them foreigners, Italians or Latins, thus implying they are not indigenous, merely the descendants of colonists.

ISLAND TOWNS: WHERE EAST MEETS WEST

During the Second World War, Hvar was under Italian occupation and then became an important military base (with Vis) after the Italian capitulation, and an important stronghold of the People's Liberation Movement in Dalmatia (Novak 1960). After the war, rebuilding began, and the population that had fled during the German occupation of the island in 1944 returned from the El-Shatt refugee camp on the Sinai Peninsula. The island stagnated until the early 1970s, when the era of mass tourism began. The influx of tourists led to huge spatial extension in urban settlements like Hvar, Stari Grad and Jelsa, while the infrastructure was built up and a large amount of migration within the island took place (Čoralić 1995). The macadam road linking the main settlements in the west and the paths in the east was only replaced by a properly surfaced road running from Hvar to Sućuraj in 1970. The eastern islanders began to move to larger urban centres for education and work, or simply in search of a better standard of living linked to the development of tourism. So, the distant, little-known "others" finally became next-door neighbours and fellow citizens. The stereotypes that used to abound between east and west were now replaced at the civic level by animosity between the "old" and "new" inhabitants, the indigenous population and the newcomers.

The town of Hvar, as an island centre and a well-known tourist resort, has many newcomers, particularly from the eastern part of the island, the mainland and the countries of former Yugoslavia. Many people from eastern villages, particularly Poljica and Zastražišće, also moved to Jelsa, while Stari Grad became a destination for people from Vrbanj, Vrisnik, Pitve and neighbouring villages, which had always gravitated toward it. We were interested to hear two types of comment. The first related to newcomers from the east to Hvar Town and Jelsa, and how they saw the citizens of these towns. The second related to the claim that newcomers had inundated the island towns and that there were few old, indigenous people of Hvar or Jelsa left.

> Mladen (Poljica, 42): "Actually, Hvar Town doesn't exist anymore. If you look in the phone book, you'll see there aren't more than 30% of real Hvar townspeople in it, so Hvar is literally isolated, on its own, and in a few years, there won't be any Hvar people left in Hvar Town. Jelsa will survive a bit longer because it is connected to the small villages around. It will still be an island place, much more than Hvar Town."

The people of Hvar and Jelsa particularly resent the newcomers because they have not adapted to or accepted the urban way of life, local speech and customs, but instead, stubbornly hang on to their own, not even trying to fit in. The established citizens emphasise their traditions and town customs, which is the main way of differentiating between "real" people of Hvar (or Jelsa) and newcomers. They also frequently mention that certain parts of the town have been taken over by newcomers (for example, Vrisak, in the town of Hvar). Although the newcomers have become citizens of the places where they live, while the eastern villages have been depopulated and only older people remain there, stereotypes still abound. They have gained a new dimension. The stereotyped "others" have moved into daily urban life, so the connotation of distance, once a reason for lumping the unknown inhabitants of the east together in one group, has been removed; however, this has not led to the eradication of division and rivalry, and has, in fact, given rise to the creation of new stereotypes based on even smaller, pickier differences. "Otherness" has simply been given a new form, and new boundaries have been drawn, this time inside the island towns. The borders between the two Hvar microregions, where "their" and "our" spaces still persist, and the new, complex border patterns show that these spaces are not homogeneous, but are continuing to branch out internally.

The eastern part of the island is still dislocated from the western part in the 21st century. It is still relatively unknown to the modern western inhabitants of Hvar, who admit that they rarely go there. The road that connects east and west, national road D116, has not been rebuilt or resurfaced since the 1970s. Zastražišće, Gdinj and Bogomolje, with their appertaining coves, are still not linked to the mains water supply.

THE MOST RESEARCHED ISLAND

Many far-reaching natural history and climate research projects were carried out on Hvar in the 19th century (Petrić 2001). Altogether, 93 researchers of various nationalities in the Austro-Hungarian Monarchy carried out the research, making Hvar the most studied Dalmatian island of the day (Mihovilović 1995).

In the early 20th century, geographer Ivo Rubić wrote about the Adriatic islands and described some noted differences between the island's inhabitants. He said that the newcomers were located in the eastern part of the island, bringing with them the influence of the "Dinaric race," a different dialect and way of life that identified them more closely with the Makarska coastal region than the "original people of Hvar" (Rubić 1952, 65, 123). Due to these divisions and dichotomies, Hvar has also been one of the most studied islands in terms of the number and variety of anthropological (particularly biological-anthropological) research projects conducted there.

Ongoing, interdisciplinary anthropological research began on Hvar in 1971 (Rudan et al. 2004) and continues today. Hvar has become a kind of model on which, during the last 45 years, according to the already established strategy for researching the island population of Croatia (Rudan 1972, 1977; Rudan et al. 1982a, 1982b, 1990), following a holistic approach, a thorough insight has been gained into the biological-anthropological and other sociocultural characteristics of the Hvar population. The noted division of the island population has been confirmed by the results of this research: quantitative traits of digito-palmar dermatoglyphs (Rudan 1975, 1977), anthropometric measures (Rudan et al. 1986), the level of kinship of assessed migration data (Jovanović 1984; Šimić and Rudan 1990), linguistic differences in basic and cultural vocabulary (Sujoldžić et al. 1982/1983; Sujoldžić et al. 1984/1985; Sujoldžić 1994) and analysis of Hvar's genetic heritage (classical antigens, HLA diversity, autosomal STRs, mitochondrial DNA and Y-chromosome polymorphisms) (Martinović et al. 1999; Barać 2002; Jeran 2010).

The complex arrangement of island borders, real and imaginary, shows that the islands are not homogeneous and that they have their own geography. The island of Hvar and its islanders show a spatial and identity-related heterogeneity that is complex and multilayered. It is the result of the island's complex history of settlement, migration and changes in rulers and systems, and its real and imagined isolation. The specific nature of Hvar as a larger Croatian island is the binary geocultural division between east and west. The western part, which has been assessed as more geographically attractive and continuously settled from prehistoric times onward, is linked in transportation terms with Split, the second-largest Croatian city.

The history of the eastern part's settlement and development has been strongly oriented toward a completely different part of the mainland—the Makarska coastal region. The links and orientation of the island toward different mainland centres have additionally influenced its divisions, based on the immigration and colonisation policies of the Venetian Republic. The division of Hvar is thus the product of an attempt to create a defensive and military border on the islands. The interaction and assimilation of newcomers and indigenous inhabitants have also been hindered by the economic-political system of benefits, which has emphasized the containment, distancing and separation of certain island regions, settlements and local communities. The border system and discourse of "otherness" have persisted for 400 years. Some stereotypes and representations of "otherness" have remained unchanged for centuries, while others have changed and even increased in response to social and cultural factors. They have also influenced the current, palpable lack of perception of the island as a whole by most islanders, and the tendency toward perceiving their own, narrow island space or dependence on the mainland.

CONCLUSION

Hvar is proof that the proximity of the mainland is no guarantee of success and modern island development, wholly or in part. Although eastern Hvar is only 7 km from the mainland and well served by ferry lines, its proximity has not made it develop more rapidly in relation to the rest of the island. On the contrary, the western part of the island, particularly the town of Hvar, is the most developed part of the island, while the municipality of Sućuraj, which covers a third of the island, is the least developed. The entire island bases its identity on the particularities and tradition of the western part and its inhabitants. All representations of Hvar islandness are built by featuring and promoting this part of the island space. The length of the island and the distance between the town of Hvar and Sućuraj (77 km), along with centuries of weak or nonexistent interisland connectivity, have influenced the creation of a marginal area in the east and the focus of the eastern population on the mainland. Statements made by islanders from the western settlements often repeat such expressions as "distant" or "far from us," highlighting actual spatial remoteness in island parameters, while some insisted that 70 km was just as far away by land or sea.

The ethnocentrism of the "first" islanders has led to the fragmentation and zoning of the island space, and the subjugation or contesting of the island identity of the inhabitants of the eastern settlements. Shaping islandness and being an islander are measured by the degree of independence from the neighbouring mainland. The essence of islandness is structured by pulling back the boundaries, emphasising the island peculiarities and contesting the right to belong to the island on the part of those who have moved there from the mainland or remain connected with it in any way. Hvar today is an example of a divided island whose strong, set boundaries permeate its interior. Often, they are harder to breach in mental and real terms than the actual island rim.

REFERENCES

Baldacchino, Godfrey. 2013. "Only Ten: Islands as Uncomfortable Fragmented Polities." In *The Political Economy of Divided Islands: Unified Geographies, Multiple Polities*, edited by Godfrey Baldacchino, 1–17. London: Palgrave Macmillan.

Barać, Lovorka. 2002. "Polimorfizam Y kromosoma u seoskim populacijama otoka Hvara." PhD dissertation, University of Zagreb.

Barbić, Jakša. 2012. *Rječnik Pitava i Zavale*, edited by Anita Celinić. Zagreb: Institut za hrvatski jezik i jezikoslovlje.

Benčić, Radoslav. 2013. *Rječnik govora grada Hvara. Forske rici i* štorije, edited by Sanja Vulić. Hvar: Muzej hvarske baštine.

Bezić-Božanić, Nevenka. 1988. *Povijest stanovništva u Visu*. Biblioteka znanstvenih djela, 29. Split: Književni krug.

Bogišić, Rafo. 1977. "Pučki i društveni odnosi u Hvarkinji Martina Benetovića." In *Matija Ivanić i njegovo doba*." edited by Josip Adamček, Radovi 10, Zagreb: Sveučilište u Zagreb—Institut za hrvatsku povijest, 419–29.

Bognar, Andrija. 1990. "Geomorfološke i inženjersko-geomorfološke osobine otoka Hvara i ekološko vrednovanje reljefa." *Geografski glasnik* 5, no. 1: 49–65

Botica, Ivan. 2005. "Prilog istraživanju najstarijega spomena vlaškog imena u hrvatskoj historiografiji." *Radovi – Zavod za hrvatsku povijest* 37: 35–46.

Božanić, Joško. 1997. "Terra mariqua." *Glasje* 4, no. 7/8: 97–100.

Braudel, Fernand. 1997. *Sredozemlje i sredozemni svijet u doba Filipa II*, Vol. 1. Zagreb: Antibarbarus.

Čače, Slobodan. 1997. "Ancient Written Sources Relating to the History of Hvar." In *The Adriatic Islands Project: The Archaeological Heritage of Hvar, Croatia*, Vol. 1, edited by Vince Gaffney, Branko Kirigin, Marinko Petrić, and Nikša Vujnović, 217–41. BAR International Series 660. Oxford, UK: BAR International.

Čolak, Nikola. 1962. "Borba novih stanovnika na srednje dalmatinskom otočju u 17. i 18. stoljeću za očuvanje stečenih privilegija." *Prilozi povijesti otoka Hvara* 2, no. 1: 52–89.

Čolak, Nikola. 1959. "Iz života iseljenika Makarske i njenog Primorja na srednje dalmatinskom otočju u 17. i 18. stoljeću." *Prilozi povijesti otoka Hvara* 1, no. 1: 87–121.

Čoralić, Lovorka. 1995. "Otok Hvar u prošlost." In *Otok Hvar*, edited by Miro Mihovilović, 72–92. Zagreb: Matica Hrvatska.

Dragičević, Ena. 1970. "Gdinjske gomile." *Periodični izvještaj Centra za zaštitu kulturne baštine Hvar* 20: 4–6.

Duboković Nadalini, Niko. 2001. "Dio otoka Hvara što se zove Plame." In *Odabrani radovi*, edited by Neda Anzulović, 501–6. Split: Književni krug.

Forenbaher, Stašo. 2002. "Prehistoric Populations of the Island of Hvar: An Overview of Archaeological Evidence." *Collegium Antropologicum* 26, no. 1: 361–78.

Forenbaher, Stašo, and Timothy Kaiser. 2008. *Grapčeva* špilja: *pretpovijesni stan, tor i obredno mjesto*. Split: Književni krug.

Forenbaher, Stašo, Timothy Kaiser, and Sheelagh Frame. 2010. "Adriatic Neolithic Mortuary Ritual at Grapčeva Cave, Croatia." *Journal of Field Archaeology* 35: 337–54.

Fortis, Alberto. [1774] 1984. *Put po Dalmaciji*, edited by Josip Bratulić, translated by Mate Maras. Zagreb: Globus.

Gaffney, Vince, Branko Kirigin, Marinko Petrić, Nikša Vujnović, and Slobodan Čače. 1997. *Archaeological Heritage of the Island of Hvar*, Vol. 1. Oxford, UK: BAR IS 660.

Gortan, Veljko. 1976. "Hvar u djelu Vinka Pribojevića." In *Hvarski zbornik*, knj. 4, edited by Milan Dorotka, 183–88. Split: Samoupravna interesna zajednica za kulturu općine Hvar, Vojna štamparija Split.

Hay, Pete. 2006. "A Phenomenology of Islands." *Islands Studies Journal* 1, no. 1: 19–42.

Jeran, Nina. 2010. "Genetic Diversity and Structure of Eastern Adriatic Islanders Revealed by Mitochondrial DNA Analysis." PhD dissertation, University of Zagreb.

Jovanović, Veljko, Branka Macarol, Derek F. Roberts, and Pavao Rudan. 1984. "Migration on the Island of Hvar." In *Migration and Mobility*, edited by Anthony J. Boyce, 143–60. London: Taylor and Francis.

Kirigin, Branko, and Božidar Slapšak. 2010. "Farska hora." In *Antički Grci na tlu Hrvatske*, edited by Jasminka Poklečki Stošić, 135–37. Zagreb: Galerija Klovićevi dvori.

Kovačec, August. 2016. "Podrijetlo i razvitak naziva Vlah, Vlasi (vlah, vlasi), vlaški." *Hrvatska revija* 3, no. 6: 3–9.

Kovačić, Joško. 1998. "Župa Zastražišće na Hvaru." *Služba Božja: liturgijsko-pastoralna revija* 38, no. 2: 159–84.

Kovačić, Joško. 1995. "Župa Bogomolje na Hvaru: o 250. obljetnici osnutka (1745–1995)." *Služba Božja: liturgijsko-pastoralna revija* 35, no. 1: 29–52.

Krasić, Stjepan. 1991. *Ivan Dominik Stratiko: Život i djelo*. Split: Književni krug.

Lajić, Ivan. 1992. *Stanovništvo dalmatinskih otoka. Povijesne i suvremene značajke depopulacije*. Zagreb: Consilium i Institut za migracije i narodnost.

Martinović, Irena, Lovorka Barać, Ivana Furač, Branka Janićijević, Milovan Kubat, Marijana Peričić, Božidarka Vidović, and Pavao Rudan. 1999. "STR Polymorphisms in the Population of the Island of Hvar." *Human Biology* 71: 341–52.

Matvejević, Predrag. 2007. *Mediteranski brevijar*. Zagreb: V.B.Z.

Mihovilović, Miro. 1995. "Pučanstvo otoka Hvara." In *Otok Hvar*, edited by Miro Mihovilović, 215–29. Zagreb: Matica hrvatska.

Milčetić, Ivan. 1912. "Dr. Julije Bajamonti i njegova djela." *Rad JAZU* 192: 92–250.

Novak, Grga. 1978. "Pismo dra Julija Bajamontija god. 1790." *Starine JAZU* 57: 75–92.

Novak, Grga. 1961. *Vis*. Knjiga prva, Zagreb: Izdavački zavod Jugoslavenske akademije znanosti i umjetnosti.

Novak, Grga. 1960. *Hvar kroz stoljeća*. Zagreb: Historijski arhiv – Hvar, Djela I, Narodni odbor općine Hvar, Izdavački zavod JAZU.

Novak, Grga. 1959. "Dalmacija godine 1775/6. gledana očima jednog suvremenika." *Starine JAZU* 49: 5–80.

Novak, Grga. 1955. *Prethistorijski Hvar*. Zagreb: Jugoslavenska akademija znanosti i umjetnosti.

Novak, Grga. 1951. "Dalmacija i Hvar u Pribojevićevo doba." In *O podrijetlu i zgodama Slavena*, edited by Vinko Pribojević, 9–47. Zagreb: JAZU.

Perinić, Ana. 2009. "Dva otoka Hvara." In *Destinacije čežnje, lokacije samoće: Uvidi u kulturu i razvojne mogućnosti hrvatskih otoka*, edited by Ines Prica and Željka Jelavić, 173–88. Zagreb: Hrvatsko etnološko društvo, Biblioteka Nova etnografija.

Perinić, Ana. 2006. "The 'Harsh Inhabitants of Hvar' in the Speech of Vinko Pribojević (A.D. 1525)." *Collegium Antropologicum* 30, no. 3: 629–35.

Perinić Lewis, Ana, and Pavao Rudan. 2012. "Spectemur agendo – prepoznaje nas se po našem djelovanju! Prosvjetiteljski rad biskupa Ivana Dominika Stratika i liječnika Julija Bajamontija na otoku Hvaru." *Rad Hrvatske akademije znanosti i umjetnosti. Razred za društvene znanosti* 49, no. 512: 57–118.

Perinić Lewis, Ana, and Pavao Rudan. 2011. "Došlo je doba svjetlosti! Stratiko je u Hvaru! – biskup prosvjetitelj Ivan Dominik Stratiko o Hvaranima." *Rad Hrvatske akademije znanosti i umjetnosti. Razred društvene znanosti* 48, no. 510: 83–100.

Petrić, Marinko. 2001. *Hvarski turistički spomenar: sentimentalno putovanje kroz turističku prošlost Hvara*. Hvar: Turistička zajednica grada Hvara.

Petrić, Nikša. 2002. "Zanimljivi crteži na pretpovijesnoj keramici Hvarske kulture: prikazi kometa iz približno 3000. g. pr. Kr." *Vjesnik Arheološkog muzeja u Zagrebu* 3, no. 35: 11–18.

Pribojević, Vinko. 1951. *O podrijetlu i zgodama Slavena*, edited by Grga Novak, translated by Veljko Gortan. Zagreb: JAZU.

Radovanović, Robert. 1976. "Iz prošlosti Gdinja." *Hvarski zbornik* 4: 287–203.

Roglić, Josip. 1977. "Prilog geografiji Hvara." In *Hvar u prirodnim znanostima (zbornik simpozija održanog 1975. g. u Hvaru u organizaciji Zavoda za povijest prirodnih, matematičkih i medicinskih znanosti JAZU)*, 5–21. Zagreb: JAZU.

Rubić, Ivo. 1952. *Naši otoci na Jadranu.* Split: Novinsko-izdavačko preduzeće Slobodna Dalmacija.

Rudan, Pavao. 1975. "The Analysis of Quantitative Dermatoglyphic Traits in the Rural Population of the Island of Hvar, Yugoslavia." *Journal of Human Evolution* 4: 585–91.

Rudan, Pavao. 1972. "Étude sur les dermatoglyphes digito-palmaires des habitants de l'ile de Hvar." PhD dissertation, Universite Paris VII.

Rudan, Pavao. 1977. "The Use of Penroses CH2 for Intra- and Inter-Population Analysis of Quantitative Dermatoglyphic Traits." *American Journal of Physical Anthropology* 46: 161–66.

Rudan, Pavao, Božidar Finka, Branka Janićijević, Veljko Jovanović, Vesna Kušec, Jasna Miličić, Marijeta Mišigoj, Derek F. Roberts, Ljerka Schmutzer, Nina Smolej Narančić, Anita Sujoldžić, Lajos Szirovicza, Diana Šimić, Petar Šimunović, and Sanja-Marina Špoljar-Vržina. 1990. *Antropološka istražavanja istočnog Jadrana, Biološka i kulturna mikrodiferencijacija seoskih populacija otoka Hvara.* Knjiga 2. Zagreb: Hrvatsko antropološko društvo.

Rudan, Pavao, Branka Janićijević, Veljko Jovanović, Jasna Miličić, Nina Smolej Narančić, Anita Sujoldžić, Lajos Szirovicza, Tatjana Škarić-Jurić, Lovorka Barać Lauc, Tomislav Lauc, Irena Martinović Klarić, Marijana Peričić, Diana Rudan, and Igor Rudan. 2004. "Holistic Anthropological Research of Hvar Islanders, Croatia—from Parish Registers to DNA Studies in 33 years." *Collegium Antropologicum* 28, suppl. 2: 319–42.

Rudan, Pavao, Derek F. Roberts, Anita Sujoldžić, Branka Macarol, Eugenija Žuškin, and Andrija Kaštelan. 1982a. "Strategy of Anthropological Research on the Island of Hvar." *Collegium Antropologicum* 6: 39–46.

Rudan, Pavao, Derek F. Roberts, Anita Sujoldžić, Branka Macarol, Nina Smolej, and Andrija Kaštelan. 1982b. "Geography, Ethnohistory and Demography of the Island of Hvar." *Collegium Antropologicum* 6: 47–67.

Rudan, Pavao, Derek F. Roberts, Branka Janićijević, Nina Smolej, Lajosz Szirovicza, and Andrija Kaštelan. 1986. "Anthropometry and the Biological Structure of the Hvar Population." *American Journal of Physical Anthropology* 70: 231–40.

Said, Edward. 1999. *Orijentalizam.* Zagreb: Konzor.

Šimić, Diana, and Pavao Rudan. 1990. "Isolation by Distance and Correlation Analysis of Distance Measures in the Study of Population Structure: Example from the Island of Hvar." *Human Biology* 62: 113–30.

Skok, Petar. 1973. *Etimologijski rječnik hrvatskoga ili srpskoga jezika, knjiga III, poni – Ž.* Zagreb: Jugoslavenska akademija znanosti i umjetnosti.

Skračić, Vladimir. 1997. "Otoci i otočani. Pogled s mula." *Glasje* 4, no. 7/8: 61–75.

Solarić, Miljenko, and Nikola Solarić. 2009. "Laying Out Land Parcels and the Oldest Boundary Stone in Croatia from the 4th Century BC." *Kartografija i geoinformacije*, 8, no. 12: 58–77.

Sujoldžić, Anita. 1994. “Govori srednjodalmatinskog otočja. Prilog antropologijskim istraživanjima.” *Društvena istraživanja* 3, no. 4–5: 423–36.

Sujoldžić, Anita, Lajosz Szirovicza, Petar Šimunović, Božidar Finka, Derek F. Roberts, and Pavao Rudan. 1984/1985. “Analiza kulturnog vokabulara otoka Hvara kao pokazatelja jezične mikroevolucije.” *Zbornik Matice srpske za filologiju i lingvistiku* 27/28: 767–80.

Sujoldžić, Anita, Lajosz Szirovicza, Petar Šimunović, Božidar Finka, Derek F. Roberts, and Pavao Rudan. 1982/1983. “Lingvističke udaljenosti na otoku Hvaru.” *Rasprave Zavoda za jezik* 8–9: 197–214.

Zaninović, Marin. 1995. “Hvar od pretpovijesti do dolaska Hrvata.” In *Otok Hvar*, edited by Miro Mihovilović, 139–65. Zagreb: Matica Hrvatska.

Chapter Nine

The Stormy Subarchipelago

The Fishermen of the Open-Sea Croatian Islands

Joško Božanić

THE SUBARCHIPELAGO

The islands, whether near or open-sea, can be divided according to whether they are uninhabited or inhabited. At first sight, the uninhabited islands may seem less interesting. The list of scientific disciplines that deals with islands with a wide range of flora and fauna, but no people, tends to consist only of the natural sciences. The list is not at all short, however, since the islands can be objects of interest to almost every natural scientist. The list for the inhabited islands is much longer, as they are frequented by natural and social scientists. They add to the complexity of island research by addressing interweaving of the island ecosystem and economic and social structures introduced by people who once reached the islands and became islanders.

This is not the sum total of island complexity as a research topic. In the world of the islands and their archipelagos there are some that are not (and often cannot be) inhabited but are still part of the economic and social systems set up by islanders. There are such islands in the Croatian archipelago. The most interesting ones are west of the island of Vis. The people of Vis call them open-sea islands, although Vis and its archipelago are in the Adriatic, a small sea in which all the islands might be called near islands in comparison with the world's oceans, or even with the rest of the Mediterranean. Why do the islanders think of them as open-sea islands and the bays that surround them as the sea, and how can a small rock protruding from the sea be of crucial importance to them? The answer should be sought in the use of island resources and a concept of the sea that dates back to antiquity and the era of the Greek colonisation of the Mediterranean.

In the 4th century BC, Greek civilisation made its way up the Adriatic. They were not interested in the interior of the eastern Adriatic and only left traces on the islands and the narrow coastal belt of Dalmatia. The island of Vis was the birthplace of urban civilisation on the eastern coast of the Adriatic. In 397 BC, Dionysus the Elder, the tyrant of Syracuse, landed there and founded Issa, the first Greek colony. The attitude the ancient Greeks had toward the sea was adopted by the Croats and has not changed to this day in the Vis subarchipelago.

There are five terms for the sea in ancient Greek: 1) *pélagos*—open sea; 2) *kólpos*—a gulf (Spanish, Italian and Portuguese—*golfo*; French—golfe; Vis [Komizan] dialect—kulaf); 3) *thálassa*—the personification of the sea—a woman made of seawater; 4) *háls*—the basic meaning is "salt"; and 5) *póntos*—a pathway. The Romans used this word for a bridge, while the Slavs transformed it into *put*, which means "path" (Vinja 1986, 38). Palagruža, the most distant island of the Vis subarchipelago, took its name from the Latin *pelagosa*, derived from the Greek *pélagos*—open sea. Situated 42 nautical miles from Vis, Palagruža is important as not only a central orientation point in the sea, but also the richest fishing grounds. It has never been easy to reach it by crossing the salty path, which is both a bay and the open sea.

In the time of Magna Graecia, when continental communication was slow and uncertain, the sea provided a faster, more regular alternative. It connected different ethnic groups and facilitated the development of large cultural communities with a sense of transethnic identity. The Croats adopted an Adriatic and Mediterranean identity from the moment they arrived from the depths of Asia and inherited an Adriatic culture. They absorbed into their language a lexicon of maritime terminology unknown to them prior.

THE DIOMEDES ROUTE

The Vis subarchipelago is a central Adriatic maritime zone whose centre comprises the island of Vis surrounded by the following open-sea islands and islets: Ravnik, Budihovac, Jabuka, Brusnik, Sveti Andrija, Biševo and Palagruža. This is the only part of the Adriatic whose geological history (some 220 million years ago) was marked by large-scale changes featuring the rise of a large mass of salt from the Earth's core toward the surface. This rising mass pushed against and pierced the upper layers of sedimentary rocks formed from petrified sand and seashells. The epicentre of this tectonic breakthrough was in wide Komiža Bay, and the peaks of the magma that rose above the sea formed the present-day islets of Jabuka and Brusnik.

The oldest, shortest and most travelled maritime route between the eastern and western coast of the Adriatic Sea begins from the Gargano Peninsula in the Italian region of Puglia. This route moves across the islands of Palagruža and Biševo, and the town of Komiža in the southwestern bay of Vis. It ends at Cape Punta Planka, also known as Diomedes' Promontory, near the coastal town of Trogir. The route was named after Trojan battle hero Diomedes, immortalized in Homer's *Iliad*. Just like Odysseus, Diomedes sailed across the Mediterranean Sea after the fall of Troy in search of new land. On his travels, he came across the Gargano Peninsula, where he founded several towns. He also explored the Adriatic Sea by sailing along the shortest trans-Adriatic route. While sailing on this route he came across a sheltered bay with freshwater springs on its beaches. He stopped there and founded a settlement, present-day Komiža on the island of Vis.

Empires and kingdoms rose and fell in the Adriatic, and the Croats who migrated there in the 7th century experienced the rule of the Romans, the Croatian kings, the Venetians and the Austrians. In the early 19th century, the British called Vis the "Gibraltar of the Adriatic," as the island was the only British trading station (*porto franco*) for commerce with Europe during Napoleon's continental blockade. In 1866, Vis was also the site of the first-ever naval battle involving steel ships (ironclads). This skirmish would lead to the defeat of the powerful Italian armada by the Austro-Hungarian fleet. The wrecks of the armada can still be seen on the Adriatic seabed. During the Second World War, the People's Liberation Army (PLA), led by Josip Broz Tito and British allies, gained control of Vis and defended the island as one of the few European territories unoccupied by German forces.

THE BLUE CORRIDOR

Based on the available scientific research revealing the exceptional continental and marine biodiversity of flora and fauna on Vis, the World Wildlife Fund included the Vis archipelago in the "Adriatic Blue Corridor" in 2003 (World Wildlife Fund 2003; Geopark Viški Arhipelag 2003). This corridor includes the islands of Sveti Andrija, Brusnik and Biševo, and the northwestern and southeastern parts of Vis, including the fishing posts around the islands of Sušac, Jabuka and Palagruža.

The Vis archipelago has unique natural phenomena (natural monuments, an ecologically preserved environment and wide biodiversity) and material and intangible cultural heritage, and it is historically important as the crossroads of maritime routes, which means it has been on the main stage of important historical events since the age of antiquity. Its natural monuments include the Green Cave (island of Ravnik), Stiniva Beach (southern coast of Vis), the Monk Seal Cave, the Blue Cave (island of Biševo) and the volcanic islets of Jabuka and Brusnik.

KOMIŽA: THE ADRIATIC CENTRE OF FISHING

The town of Komiža used to be the centre of the fishing industry on the east coast of the Adriatic. At some point, the fishing grounds became too restricted and perhaps not challenging enough, and the people of Komiža started fishing beyond the Adriatic. They reached the Atlantic in the early 20th century and built barracks for salting fish in Costa da Morte on the Spanish Galician coast (Finisterre), contributing to Spain's early fishing industry. The next stop was the Pacific, where Komiža fishermen joined the pioneers of modern fishing from Chile to Alaska: The California town of San Pedro became a fishing centre due to its Komiža immigrants. Back home, the architecture of the historic urban core of Komiža reflected the fishing community in a unique way. All the buildings by the sea were small family workshops for salting fish. The Komiža fishermen produced a unique type of fishing boat, the *gajeta falkuša*, designed for the specific, often-extreme weather conditions in the open-sea islands around Vis. It was in use until the mid-20th century.

It was the *falkuša* that enabled the people of Vis to become and remain the only open-sea fishermen in the Mediterranean until the mid-19th century. One of the oldest canneries in the Mediterranean, the Mardešić factory, was founded in Komiža in 1870. In the early 20th century, Komiža fishermen made a significant contribution to the development of the fishing industry in the United States, too. A native of Biševo, Martin Bogdanovich, founded Star-Kist, one of the first fish canneries on the West Coast of the United States, which has since become one of the largest fish canning factories there.

LIBER COMISAE

Written sources about fishing traditions on Vis in the Greek and Roman eras are scarce. There have been archaeological finds, for example, sinker weights and iron fish-hooks; however, as the vibrant fishing tradition of Komiža has not faded away, the main relics still can be seen in the harbour. A couple of *falkuša* with *vojga* gill nets are always ready to sail out.

The most valuable source is the *Liber Comisae*, a unique collection of manuscripts containing records of the fishing history of Komiža. It was compiled by Nikola Borčić Jerolimov in the second half of the 18th century. Today, the manuscripts are kept in the parish archives in the town of Vis. They contain descriptions of the fishing communities from the 16th to the 18th century and cover turbulent periods marked by intense conflicts between the fishermen and the local authorities.

The oldest document in the *Liber Comisae* contains a description of a regatta from Komiža to Palagruža, a boat race in which the contestants claimed the best posts on the pebble beaches of Palagruža by pulling their boats ashore. In the Town of Hvar, which was then the administrative centre for Vis, 74 *falkuša* owners were officially registered on 9 May 1593, and took part in the regatta during the next new moon (Novak 1952). Other sources reveal that the fishermen who wanted to participate in the regatta had to report to the head captains, Franjo Borčić and Jakov Bogdan, on 20 May. Unregistered boats were strictly prohibited from participating in the regatta, and those who disobeyed this rule were arrested and sentenced to work as galley slaves for 18 months. The sources further tell us that these regattas were escorted by a war galleon as additional protection from pirate attacks. The duke of Hvar ordered the fishermen to arm themselves with harquebuses and forbade anyone younger than 18 years of age to participate. The fishermen had to obey the orders of the head captains, unless they wanted to risk being jailed in the Hvar prison (Novak 1952).

PALAGRUŽA: AN OPEN-SEA ISLAND

The story of Palagruža occupies a prominent place in the history of Croatian fishing. For centuries, Komiža fishermen have sailed on *falkuša* to the island's fishing posts. Pilgrims on their way to the Holy Land have described seeing fleets of boats manned by Komiža fishermen heading for the open-sea island of Palagruža (Kužić 2013, 137). This island was a rich fishing zone, and the fishermen of Komiža caught large quantities of blue pelagic fish, which were salted and exported to Venice and the countries of the Near East.

Aside from extreme weather conditions, the open sea was also dangerous because of the constant threat of pirate attacks. For this reason, the Komiža fishermen always sailed in large groups to Palagruža. These expeditions consisted of as many as 100 *falkuša* boats, and their joint forces were enough to fight off pirates who went after these fishermen to not only steal barrels of salted sardines, but also take strong men into captivity and sell them later as galley slaves.

The first documented regatta took place in 1593. In the early morning of 20 May, the cannons on top of Komiža tower went off, and the rumbling sounds began an armada of 74 *falkuša* boats headed by a war galleon (Novak 1952). This procession set out toward the island of Palagruža, 42 nautical miles south of Komiža. There were five fishermen aboard each *falkuša*, and since there was no wind in the sails, some 370 oars were manned and deployed, churning up the sea. The fishermen had to row for as many as 15 hours to reach the beaches of Palagruža.

The men of Komiža made this long voyage to reach the rich fishing grounds and also because the fishing posts around Vis and Biševo were already occupied by fishermen from the neighbouring islands of Hvar and Brač, who used large seine nets, while the Komiža fishermen used smaller gill nets for catching pilchard. The fishing positions around Vis and Biševo were convenient for the people of Hvar and Brač, and there were four times as many of these large teams, each comprising 15 to 20 fishermen on three to four boats.

Seine nets were costly to make and maintain, and the owners of these boats, especially those from the island of Hvar, were affluent and on good terms with the local authorities. Most of them did not take part in the fishing expeditions, but they nonetheless looked down upon gillnetter boat owners. They managed to persuade the local authorities to impose fishing restrictions on gillnetter boats, eventually implementing a total ban, with drastic measures taken against transgressors. Their motive was to eliminate competition at sea, since pilchard caught by gillnetter boats were of better quality and commanded higher prices. Moreover, the owners of the seine boats needed larger crews and wanted the gillnetters, who were highly valued as strong oarsmen and for their expertise, to work for them. When fish were scarce, the gillnetters were allowed to fish in the vicinity of the seine boats; however, if fish were plentiful, the gillnetters were forbidden to fish, even around the waters of Palagruža, where the underwater and coastal terrain was unsuitable for seine fishing anyway. Despite the draconian laws prescribed by Venice and implemented by the state judges of the Hvar commune, the fishermen resisted and continued their fishing expeditions at the risk of losing property and life.

DRACONIAN PUNISHMENT

On 25 May 1655, a sentence was passed in absentia on 20 fishermen from Komiža accused of pilchard fishing around Palagruža. Part of the verdict read,

> Conscious of their gross transgression, they dare not return to face justice. To punish such outrageous insubordination, we pronounce that the absent subjects Bartul Zamberlin, Petar Petričević, Andrija Harvat, Petar Covo, Mihovil Pavšić, Pavao Bracan Jr., Franjo Fruzin, Jakov Galiot and Ivan Perina be exiled from this town and region, and from the Province of Dalmatia, for a period of 10 consecutive years, and that they should approach no nearer than the tower of (the coastal town of) Klis. If during that time they are found in defiance of this ban and brought to justice, let them serve for five consecutive years as oarsmen,

> chained by their feet, on a galley of *La Serenissima*. Should any of them be incapable of such service, he shall be kept in jail for three years and then returned to exile. . . . Anyone who catches or kills these offenders shall receive an award of 100 liras for each man, and this money will be taken from the offender's property. If he has no property, one half of the amount will be secured from the funds of this municipality. (D'Erco 1973, 192–93)

Drastic measures were constantly invoked: Fishermen were exiled from their homeland, sentenced to years of slavery rowing on galleys bound in chains, and their property was confiscated and their nets burnt. The notice of the general overseer, Alvis Mocenigo, dated 1697, regarding the ban on gill net fishing demanded "burning of their boats and nets, fines of 50 ducats per net, exile, prison and hard labour" (D'Erco 1973, 194).

In 1673, the Komiža fishermen decided to address the doge of Venice directly and wrote a petition. It was written on behalf of them all and signed on 16 May 1637, by the local procurators, Antun Delić and Vinko Karuza, and Antun Maffio and Vid Bogdan, the head captains for fishing expeditions to Palagruža. In the petition, the people of Komiža pointed out that owners of gillnetter and seine boats had once enjoyed the same rights and competed for the best fishing positions. When the gillnetter boats lost those rights, they were also forbidden to fish around Palagruža, despite the fact that seine fishing was impossible in this location. They also pointed out that *falkuša* boats travelled to Palagruža in large numbers as a defence tactic against pirates. The fishermen were praised for looking out for and informing the navy and Venetian authorities about enemy ships, since Palagruža was closer to the Italian coast and Apulia than Vis. They were also commended for chasing Apulian fishermen off Palagruža (Novak 1952).

In 1698, the Komiža gill net fishermen composed another petition addressed to the doge of Venice. This time, they asked for permission to engage freely in pilchard fishing, since Komiža was a "part of the island of Vis which is also the most barren and infertile part of Dalmatia, with rocky terrain and bare hills." They continued,

> Nothing can grow there because it lies in the jaws of the sea, and its fertility is no greater than that in Arabia. Komiža could be mistaken for Arabia, were it not for the sea, which feeds about 1,800 hapless souls . . . who survive on income from pilchard fishing, using smaller gill nets rather than the larger seine nets, which they cannot afford. (Novak 1952, 196)

The fishermen reminded the doge how valuable Komiža fishermen had proved to Venice during its wars with the Ottoman Empire and complained of the harsh and unjust penalties they endured during the gill net ban. They stated that their survival depended upon fishing around the remote islands of

Sušac and Palagruža, where they faced great danger from pirates who had "taken captive 176 of their fellow fishermen in several raids" (Novak 1952, 196).

Yet another letter for the doge was sent to the Council of the Wise in Venice in July 1696. In 1700, the petitioners were visited by Venetian warships, and their nets were burnt; however, the fishermen did not relent. Ricardo D'Erco, in *History of Legislation on Sea Fishing in the Waters of the Austrian Littoral* (mid-19th century), wrote the following:

> The Winged Lion raged against them constantly with all the force of its wounded authority, except for a few moments of truce. It persecuted them incessantly by confiscation, burning and threats of jail sentences, exile, hard labour and even death. Children lost their fathers, wives their husbands, and families were deprived of their sole breadwinner. But such cruelty led nowhere. As a hydra grows another, when its head has been cut off, so the confiscated and burnt gillnetter boats were replaced with new ones. The gill net fishermen withstood Venice's rage and remained untamed and undefeated in a ferocious struggle with the Venetian Republic over a period of 200 years. (D'Erco 1973, 183)

THE ISLAND OF SVETI ANDRIJA

Looking westward from Komiža Bay, we can see the contours of an open-sea island on the horizon. This is Sveti Andrija, until recently the most distant inhabited island in the Croatian archipelago. It is named after the patron saint of fishermen, the apostle Andrew. Here, at the gates of the highly indented Mediterranean archipelago, with more than 1,200 islands, it rises 311 metres above the sea. The island is uniquely solitary and inaccessible. No tourist routes lead to it. Even a determined tourist would find it impossible to land on its rough, exposed shore. Aside from the Zanki family, who occupied the island for two centuries, only occasional poets or painters have lived here for a while, far from the madding crowd. Komiža fishermen used to visit the island frequently and use it as a refuge.

In his renowned travelogue *Viaggio in Dalmazia*, published in Venice in 1774, Alberto Fortis claimed that the island called Pityeia in *Jason and the Argonauts* by Greek epic poet Appolonius Rhodius (3rd century BC) actually refers to the island of Sveti Andrija (Fortis 1984). This open-sea island is located between the two Adriatic coasts and has been inhabited since prehistoric times. Archaeological finds in its caves include stone axe-heads, daggers, fishbone combs, lead anchors and amphorae retrieved from the seabed around the island. They indicate the importance of this location along the maritime route between the two coasts. Artefacts found at underwater sites tell a story of sea journeys in Greek and Roman times that ended on its harsh cliffs (Kirigin and Milošević 1981).

The northeastern peak of Sveti Andrija above Cape Sijabod is known in the local oral tradition as *Krajicino*. Archaeologists have determined that a fortification was erected there in the 6th century as a control point on the maritime route between the two coasts (Kirigin and Milošević 1981); however, according to a prevalent, often-cited legend in Komiža's oral tradition, this hilltop fortress was erected by Illyrian queen Teuta, whose fleet held the well-fortified town of Issa on the island of Vis under siege for many years in the 3rd century BC. Teuta is referred to as the mysterious "queen of the Islands," *Krajica ol Otuk* in the archaic speech of Komiža and Sveti Andrija. According to the legend, Teuta monitored the sea traffic in these waters from this fortress. If she spotted a boat, she would quickly send her fleet to intercept it, rob it and take the sailors captive as her sex slaves. They would ultimately meet a bitter end, as she would throw them from the walls of the fortress down the cliff when she tired of them, so the story goes.

Resin Collectors

In 1760, the affluent Foretić family from the island of Biševo organized a small resin factory on the island of Sveti Andrija, which was listed as an actual industrial object (*la fabbrica della Pegola Spagna*) of island enterprise (Novak 1952). Tars made through the dry distillation of wood and raw pine resin were highly valued products for the Venetian shipyards, where they were used to seal boat hulls. To begin this enterprise, the Foretić family brought some 20 *pegajuoli*, resin collectors from the town of Peschici on the Gargano Peninsula, to Sveti Andrija. The same document shows that the island products were highly sought after because they were of higher quality than tar and resin from the Kingdom of Naples (*per opinione de' Corispondenti erano considerate di qvalità migliori di qvelle dal Regno di Napoli*); however, since the forests of the island were exploited beyond any sustainable level, they were soon depleted and devastated, and the factory had to shut down.

MEMORIES

The late Ivan Vitaljić Gusla spent a good part of his life fishing in the waters around the islands of Svetac and Brusnik, catching rock lobster in winter and pilchards in summer. He spent many days and nights in Mancinovica Cave on Sveti Andrija. I recorded Ivan's story on a lovely summer day in front of his house in Komiža on Kalafotova Beach in 1983.

> I spent many years fishing around Sveti Andrija in winter, catching rock lobster with traps. The Komiža fishermen used to catch lobster with gill nets, but they switched to lobster traps after watching an Italian fisherman using them to catch lobster on Sveti Andrija. His name was Nicolo, and he was from Puglia. While fishing around Sveti Andrija, he would live in Mancinova Cave in Zaruška Cove, where you can pull your boat ashore when north winds are strong, and sometimes when the south winds blow. He taught the Komiža fishermen how to catch lobster with traps.

Living on this island with no safe dock or shelter was not safe or comfortable. In a sense, living meant sailing. It meant always staying on the alert, even in your sleep, and straining your ears for the slightest change in weather, the tiniest puff of wind, which could gain enough strength in an hour to raise mighty waves whose roaring and menacing thunder could shake the entire island.

The boats were not safe even when pulled ashore, because the sea could stretch its mighty hands for 100 metres up the slipway. I recorded a story Ivan told me that vividly evokes the memory of a terrible event from long ago, when the names of those involved had faded from memory:

> A stormy southern wind picked up at night, and people ran to pull their boats, already on the slipway, further up the slope and as far away as possible from the sea. But a large wave broke through a side channel behind them, sweeping away both people and boats. Only one woman survived the ordeal because her skirts got caught on the stone mooring post on the slipway. And so the sea swallowed the men and their boats, and left no trace, except for the story, as a reminder to all those who do not believe how powerful the sea can be.

THE ISLAND OF JABUKA

Rising from the sea like a black pyramid, the island of Jabuka is some 30 nautical miles west of Komiža. In 1958, it was declared a geological natural monument. It is 97 metres high and roughly 700 metres in circumference. Jabuka is composed of deep-crust magma rock, whose magnetites interfere with seafarers' compasses, making navigation extremely difficult in conditions of poor visibility. An endemic carnation (*Dianthus multinervis*) grows on the top of the island.

Our storyteller, Komiža fisherman Ivan Vitaljić Gusla, told us a story about Jabuka. As the tape rolls, Ivan's voice relates his story. He is no longer with us, so now we have only his voice:

> Jabuka was usually shrouded in fog, so you could not see it. Even if you rowed in calm waters still it was difficult to see it. It was difficult to spot it even if

you came really close. You would sail along a maestral wind and often think that you had probably gone far enough to reach it, but the island was still nowhere to be seen. Then you would usually stop. You would know by the sun what time it was and how much time you had been sailing. After three to four hours of sailing you should have been able to see it. You reckoned the time by the sun. There was no clock. We would measure the sun's position with our fingers, so we could tell the time. And the shadow of the mast at midday could also tell you what time it was. You would measure against the western horizon with your fingers, to see how high the sun was. I remember there were times when we were only half a mile away from Jabuka but could still not see it. But when the seagulls from Jabuka sensed you coming, they would start to squawk, so we could locate Jabuka by their sound and then finally see it. Or if there was some swelling at sea, we would stop and listen carefully for the sound of waves crashing against the shore. In the springtime, when the seagulls had their young, the adult gulls would fly out to meet us fishermen and we would throw them some fish, and they would fly with the fish in their beaks straight towards Jabuka to feed the fledglings in the nests. You could tell by their flight route where Jabuka was hiding in the fog. The compass was of no avail near Jabuka. Too many rocks full of iron.

Crosses on Jabuka

Now imagine the life we led out on the open seas where you can see no end to it, a small island without a dock in the middle of nowhere. There are only two places where you can disembark, and you have to be young and strong to get out of the boat.

When I first arrived on Jabuka, under its cliffs, I looked up and it seemed to me that the rocks would come crashing down on us. There was nothing to hold on to, everything was slippery. I said to myself, "What in the world is this place, where am I?" When I spotted four crosses drawn in tar on a flat rock, it sent tingles down my spine. The older fishermen then told me that on that very spot, four fishermen from the island of *Žirje* had died.

Conger Instead of an Alarm

In most cases, the captain kept watch overnight, because he was afraid one of the crew would fall asleep from exhaustion. He trusted himself best, because he was in charge, along with the oldest member of the crew. He would sit on the stern and look out for any change in weather. If the weather was stable, with a light northern tramontana wind and the stable northwest maestral in the afternoon, the captain would still keep watch, but with less anxiety. But if there were calm seas, which foreshadow change, then everyone was tense and on the lookout, for fear of being caught off guard while sleeping. A strong storm could arise suddenly overnight, or a strong north wind. And so the weary crew would fall asleep, and the captain would cast the line for conger to keep him alert and awake.

In the morning, the captain would make tea and shout to the crew: "Ala, my men, let's go!" But waking up was a terrible ordeal! Hands and fingers had swollen overnight. If you wanted to take a leak in the sea, it was difficult to unbutton your fly. Believe me, there were times when you could not unbutton your fly in the morning. So the men would mostly unbutton their flies the night before, while they could still move their fingers. And as you were taking a leak in the morning you would wet your hands with your own urine so your hands would soften, and the swelling in your fingers could die down and you could start working. The most difficult ordeal was to lift the first two buoys before your fingers warmed up. Our hands would get injured by the lobsters. Then you would have to pack them in a box on seaweed, and each lobster had to be repacked several times so it could let its excrement out. If you were on the island for three days, you would have to repack the same lobsters three times. You would handle the same lobster three or four times, and lobsters are rough. The boards were always wet, and you would have to sleep on those boards.

Carnations on the Volcano

Jabuka is so lovely in the spring, there are thousands of seagulls that live there, and they keep the islet fertile. There is also wild clover. Jabuka is lush in the spring. The black island is like one big bouquet of flowers in the middle of the sea.

Our old folk used to believe in fairies. The old captains would used to say to the young men while they were fishing on Jabuka, "Go, lads, to the very end. Climb up Jabuka and pick some spruce so we can light a fire, but do not climb to the very top and do not pluck the carnations that grow on the top. The fairies grow those carnations. They plant them, fertilize them and water them. In the spring, the fairies come here from Mount Velebit to swim and they are the ones who grow these carnations. Don't touch them and don't pick them. Anyone who picks them will be caught in a storm on Jabuka." But the young men nevertheless secretly picked the carnations so they could give them to their girlfriends back home. Because these were carnations from Jabuka! If they got caught in bad weather on Jabuka, the older folk would say, "That happened because of the carnations; one of our younger men must have picked them!" They were not just afraid of bad weather, but also that Jabuka's rich fish stocks might disappear, because they were their sustenance and their families' source of income. The older fishermen would warn the younger men, "Lads, don't pick the carnations on Jabuka! As long as there are carnations, there will be fish as well!"

THE ISLAND OF BRUSNIK

Eleven nautical miles from Komiža and three nautical miles from Sveti Andrija lies the small islet of Brusnik. It was declared a geological natural monument in 1951. The island has Paleolithic pebble conglomerates on cliffs high above sea level. There is a ravine with a depression in the middle of the islet filled with seawater, where fishermen from Komiža made small saltwater lobster pools using large pebbles. On its large black pebble beach there are remnants of fishing huts built from large stones and covered with pine branches from the nearby island of Sveti Andrija. I listened to a story from Ivan's tape, recorded in his barrack in Komiža in 1983:

> During harsh winter storms, the swollen sea would split the island of Brusnik into two smaller islets. The sea would break through the central ravine, Dolac, and fill the space so a *gajeta* could sail freely through the middle of the island.
>
> Brusnik is like a large boat anchored in the middle of the open sea. And just imagine all those young men there. The captains were older, but the crews were mostly younger men, 20, 25, 30 years old at the most. When a fisherman turned 40, he was no longer capable of rowing. That required strength and endurance.
>
> Even the seagulls would leave Brusnik in winter. Only black lizards and fishermen remained. If the weather was nice, we'd go lobster fishing. The captain would be on the prow giving commands to the oarsmen, how to row, how to turn opposite the current so the traps would not get stuck to the bottom. We were not supposed to lose any traps.
>
> There were lobster pools in Komiža owned by the Marinković family that could accommodate up to a wagon of lobsters. From there they would export lobsters all over Europe. Lobsters would only be sold at Christmas and Easter. Live lobsters were packed in wooden boxes on dry seaweed and shipped to Trieste, Paris, Rome, Prague, Milan and Vienna. No one in Yugoslavia knew about lobsters then. And the owners of the lobster pool in Komiža, the Marinković family, exported up to two tonnes of live lobsters to European cities in one go. Meanwhile, Komiža folk ate only garfish and pickerel in winter. This was the fare of the ordinary folk.
>
> And so, there was all this hard work, and swollen hands, and sleeping with unbuttoned flies so you could take a leak in the morning. You couldn't tell that to anyone today, because they would say you were crazy, that it was impossible. But it was really like that, because you're not hearing this from someone who has heard it from another but from someone who lived through it, who cried in excruciating pain in the morning when he had to pick up a rope again to start lifting the lobster traps. We peed on our hands so we could get those fingers to move, so that ammonia from the warm urine would relax them. That was our life. Our hands were hardened like donkeys' hooves from handling all that rope.

And so this life rolled on, year after year. Only the memories remain, though we old folk have no one to tell anymore anyway. But as difficult and as hard as our lives were, we led joyful, merry lives. There was more friendship, more mutual respect, trust between people. People were not indifferent to each other. It was just, how shall I put it, a different approach to life altogether.

THE ISLAND OF BIŠEVO

The island of Biševo has a surface area of 5.92 km², and its coastline is approximately 18 km long. Biševo's highest hill, Stražbenica, rises above the large southern bay of Trešjavac to a height of 239 m. Biševo has eight villages: Mezuporat, Poje, Potok, Velo Gora, Nevaja, Porat, Sarbunora and Gurnjo Sarbunora. As many as 320 people lived in these villages before the First World War. They were all vine-growers and fishermen, and each family had a home in Komiža. Biševo was once a vine-growers' island covered by large vineyards planted in fertile, sandy soil. A renowned black grape variety known as *Biševo plavac* was grown here, producing superior-quality wine. It is said it was even served at the Imperial Palace in Vienna. Having vine growing and fishing well balanced, Biševo was probably the best example of the division of labour in the archipelago. Fishing expeditions took more than a day, sometimes a couple of days in the sea; work was hard and there was a lot of rowing. There was a common understanding that only men could do that and also enough superstition about women and the bad luck they bring if on board. They stayed ashore, attended vineyards, took care of the household and waited for the men to come back with the fish.

Fishing spots along the coast of Biševo were rich in pelagic fish, which fishermen caught in large quantities. They salted pelagic fish and exported it to Venice, Southern Italy and Greece. It is said that the fishing taxes from the bountiful catches from Biševo's best fishing spot, Trešjavac, were used to build Komiža's Komuna fortress and complete St Stephen's Cathedral in the town of Hvar.

THE BIŠEVO FISHERMEN WHO LEFT THE DOOR OF THE PACIFIC AJAR

The emigration waves from the Croatian islands did not bypass Komiža, Biševo and Sveti Andrija. Fishermen from the subarchipelago went as far as the West Coast of the United States and continued what they were best at: fishing. There were many of them, but two were exceptional. Martin Bogdanovich (1882 [Biševo]–1944 [San Pedro, California]) arrived in San Diego,

California, in 1911, and became a fisherman. He made his living from trading fish and opened his own factory in 1917. The story goes that he started using crushed ice to preserve the fish at sea, a necessary component of deep-sea fishing far from the shore. The strength of his cannery was based on the exceptional skills and experiences of people from his homeland. Bogdanovich went from a poor country to a land whose West Coast offered opportunities to those who were competent and courageous. The poor fisherman of Biševo made a name in American fishing history.

The other prominent character was Paul Martinis (1893 [Biševo]–1974 [Everett, Washington]), who was born on the island of Biševo in 1893. Instead of doing his military service in the Austro-Hungarian army, he decided to go to the United States in 1913. He boarded a steamship in Trieste and, after an 18-day journey, disembarked in New York without knowing a word of English and with only $22 in his pocket; however, he drew upon his fishing experiences with his father in his early years. Hearing that there were fishermen from Komiža in Tacoma, Washington, he set off in search of work. It was a pioneering time in fishing in Alaska, when fishing was still powered by sail and oar. Paul Martinis and his brothers, who left Biševo to join him in the United States, went down in the history of American fishing not only because of their record catches of salmon in Bristol Bay and the Bering Sea, but also as the first American fisherman who mastered salmon fishing in the rich but dangerous waters of the Aleutian Islands. The fishing community of Everett, Washington, where he spent his last years, nicknamed him the "King of Salmon."

WHAT HAPPENED, AND WHAT IS TO BE DONE?

Whatever happened to the people of these Croatian islands, insular posts with intense maritime traffic, where different Mediterranean cultures exchanged languages and experiences? Whatever happened to the hardworking, resistant, enduring fishermen who for centuries lived with and from the sea? What happened to their language, which reflected centuries of hard toil and survival on the crossroads of maritime routes? Those men possessed a fine sense of balance in their relationship with nature and an awareness that they were part of the circle of life. The subarchipelago they lived on for centuries was a microcosm of independent, strong people who perfected a whole set of heterogeneous skills necessary for survival. When great global changes reached those shores in the mid-20th century, the old, familiar world suddenly became outdated. There was no place for it in the new order of things. So intense were these changes that the islanders could no longer keep up with or control them. And the changes have definitely disrupted the precarious balance of life practised and sustained for centuries.

Homo insularis did not adapt well to the new rules of the new age. There is no one on Sveti Andrija today, and hardly anyone lives on Biševo. The open-sea subarchipelago of the Adriatic is desolate, and the places underneath and above the sea are devastated. The question remains: What is be done?

REFERENCES

Borčić Jerolimov, Nikola. 1784. *Liber Comisae conscriptus anno Reparatae salutis 1784*. Manuscript from the 18th century kept in the parish office in the town of Vis (island of Vis).

D'Erco, Ricardo. 1973. *O ribolovu na istočnom Jadranu*. Zagreb: Jadranski institut JAZU.

Fortis, Alberto. 1984. *Put po Dalmaciji*. Zagreb: Globus.

Geopark Viški Arhipelag. 2003. "Viški Arhipelag—Adriatic Blue Corridor." *Geopark Viški Arhipelag*, https://geopark-vis.com/bastina-vis/prirodna-bastina/viski-arhipelag (accessed 11 June 2020).

Kirigin, Branko, and Ante Milošević. 1981. "Svetac." *Arheo, Arheološka obvestila, Glasilo slovenskega arheološkega društva* 2: 45–51.

Kužić, Krešimir. 2013. *Hrvatska obala u putopisima njemačkih hodočasnika* 14–17. *st*. Split: Književni krug.

Novak, Grga. 1952. *Dokumenti za povijest ribarstva na istočnoj obali Jadranskoga mora: sv. 1, otoci Vis i Hvar*. Zagreb: JAZU.

Vinja, Vojmir. 1986. *Jadranska fauna: etimologija i struktura naziva*. Vol. I. Split: Logos.

World Wildlife Fund. 2003. http://worldwildlife.org (accessed 6 February 2020).

Chapter Ten

On the Rim of Croatia and Croatian Development Policies

Željka Kordej-De Villa and Nenad Starc

"Each house of the people who have land on the island of Lastovo may keep 60 small animals and no more. . . . In the same way, each house may have oxen, cattle or beasts of burden, eight of them, and no more" (*Lastovski statut* 1994). This strict regulation was included in the ancient Statute of the Island of Lastovo, which was once part of the Republic of Dubrovnik, in 1390. It was not a large island, a mere 38.60 km^2, but it had 46 fertile karst fields that were cultivated to their full extent. When livestock began to threaten sustainable agriculture, the main island regulatory document decreed that their numbers should be controlled. The carrying capacity of the ecosystem is difficult to determine even today, and we do not know how the people of Lastovo and Dubrovnik arrived at the figure of 60 small and eight large animals, but it is clear that 600 years ago, the most distant island in the Republic knew how to manage its resources and institutionalise them.

The statutes of Krk, Korčula, Pag, Rab and other islands show that these things mattered throughout the archipelago. The decrees and instructions issued by the coastal city of Zadar to the islands in its subarchipelago also show that the island tenants cared about sustainability and engaged in agriculture, crafts and fishing, giving a due part of their production to the landlords who owned the islands (see Chapter 6). The islanders had to produce enough for themselves and their landlords, with the latter being careful as to not procure too much so that there would be something left for the next year. The sustainable use of resources always went hand in hand with sustainably exploiting those who cultivated the resources. Wherever and whenever possible, sustainability was institutionalised.

In the previous chapters, the ageing of the island population has been highlighted, along with the demographic atrophy of the small islands, unlicensed building, the uncontrolled growth of tourism and, above all, deinsularisation with the accompanying disappearance of islandness. The new Islands Act has been mentioned and even commended, but the question of what is to be done has been posed nevertheless. In this chapter, we will consider island developmental policies. This has a long history, as it refers to near islands that have always been under the dominion of a mainland empire, kingdom or republic, so that decisions on island development have always been made on the mainland. A chapter devoted to this topic is necessary, as the Republic of Croatia is one of the few countries in the world that has had a special island development strategy in the past quarter-century and a special Islands Act. So, it needs to be seen why, alongside wider experiences in managing island development, key developmental issues are becoming more frequent and more urgent.

TENANTS AND LANDLORDS

Sustainable exploitation of those who used the island resources for centuries was the guiding principle of island economic policies in the Adriatic. These policies were conceived in far-off Constantinople, later in somewhat nearer Venice and even later in Vienna, but the local hierarchy was charged with their implementation. Yet, they were imposed top-down on the Croatian population who had arrived on the islands 1,300 years earlier and stayed. Throughout that time, they were subjects, even of their Croatian rulers. From the late 9th to the late 11th century, the geopolitical situation and resources of the islands were observed and assessed in the mainland towns of Nin, Biograd and Knin, where the Croatian dukes and kings had their seats. The south Dalmatian islands of Lastovo and Mljet, the Elafiti subarchipelago and the Pelješac peninsula were part of the Republic of Dubrovnik for hundreds of years, and their resources and tributes were managed from Dubrovnik right up to the early 19th century.

From the Kvarner islands in the north to the Dubrovnik islands in the south, the socioeconomic system was marked by the *colonat* system (see Chapters 3 and 6). Economic and development policies, in general, were handled by both sides; the tenants were interested in managing their resources sustainably and as productively as possible, while the landlords and their overseers maximized tributes to ensure that the exploitation of tenants remained sustainable. The annual cycle of the island economies continued without significant changes for centuries. It would start with an assessment of resources and potentials, and continue with an estimate of production levels (usually overestimated by the landlords' overseers, and deliberately underestimated by the tenants). Long months would pass, punctuated by occasional inspections. At harvest and grape-picking times, the despised overseers from the mainland, who were only slightly less loathed than the marauding pirates from the sea, would arrive to collect their share of the produce and take it to the mainland or the landlords' island summer residences.

Under this system, production improved, and productivity grew sporadically, depending mostly on the care and efforts of the islanders. Innovations were rare, even on the mainland and, in any case, tended not to reach the islands. If we were to define the economic policy, as economists usually do, and wonder who its bearers and instruments were, what they had at their disposal, what measures they undertook and to what end, we would find only rudiments of what is today understood as development management. Instruments of force were wielded often and felt when something was either prohibited or claimed as a tribute. On the other hand, such incentives as reduced tributes or ceding land to the tenants were rare. The tenants could have left the islands if they had wanted to, but few dared risk such an undertaking. Little news filtered through to them of other places where life was perhaps better.

The Venetians on the other side of the Adriatic did not offer any better conditions than they already had on the islands. They went to the mainland towns as day labourers but were not allowed to settle there, and the hinterland was mostly under the control of the Ottomans, about whom they had only heard the most terrible things. Accordingly, the near islands were managed from the mainland and on the islands. Tribute levels were decided on the mainland, although production was decided by both sides. Islanders were better informed of island resources and subsequent yields, and landlords knew which products could be sold on the market and at what prices; however, the organisation of production, productivity and quality was left almost entirely to the islanders. As time went on, overseers who collected dues on behalf of landlords learned to estimate annual wine and olive oil storage, and gauge lime surpluses and the amounts of fish in need of salting. But they only had a vague idea of how these items had got there and whether there could have been more. Most landlords, who visited the islands only to hunt or stay in summer residences, hardly ever got involved.

The *colonat* economics of near islands reveal a little more freedom than the tenant–landlord systems on the mainland, particularly when compared to serfdom on the mainland and interior. Insularity did not affect the functioning of the feudal system, but it forced landlords to allow their island tenants more autonomy. Although the islands were visible to the naked eye and could be easily rowed to and controlled militarily, hardly anyone interfered with island production. The goals of economic policies imposed by landlords at that time were met by simply depriving islanders of part of their produce. The efficient use of limited resources was soon saturated, while the imperative of sustainable exploitation determined the optimal structure of production and its highest possible levels.

The island was the *locus* for use and exploitation, but its insularity and scarcity of resources required that landlords, who were not known for benevolence despite their limited understanding of these resources, adhere to sustainable methods. They realised that the islands could not be successfully exploited if they were not seen as foci and accepted on their own terms.

MAINLAND-MADE ISLAND POLICY: SETTING THE STAGE

The *colonat* system on the Adriatic coast and islands lasted as long as the sociopolitical systems of the states that exploited them. As the ruling states often changed in the late 18th and early 19th centuries, the tenant system was relaxed and reimposed in turn, so that on some islands and parts of the mainland it was only abolished after the Second World War. The news of Napoleon's troops taking Venice in 1797, abolishing La Serenissima in 1805 and ceding it to Austria slowly filtered through to the islanders. It was only in 1866 that Venice became part of the Kingdom of Italy, whose borders extended into and ended in Istria. This ended the Venetian domination of the eastern Adriatic and most of the islands, which had lasted from the 11th century with occasional interruptions and many conflicts, repossessions, and repurchases.

The news also reached the southern islands under the Republic of Dubrovnik. In 1806, La Grande Armee took Dubrovnik and, two years later, abolished the Republic. They annexed it to their Italian kingdom, and then to the Illyrian Provinces, where it stayed until 1814. In 1815, after the Battle of Waterloo and Congress of Vienna, the eastern Adriatic up to the Bay of Kotor and all appertaining islands were annexed to the Austrian monarchy. It was reorganised in 1867, into the dual Austro-Hungarian Monarchy, so rather different flags flew on the islands and official documents acquired extensive letterheads. Economic policies stayed the same, on the whole, for the islands and the mainland. The monarchy maintained them until 1918.

These changes, which caused much turmoil in Europe, are worth enumerating, as the islands that could be reached easily from the mainland observed the bloodshed from the front row, without being involved themselves. They were closer to the battlefields than many European capitals, but most of them were strategically unimportant and remained in the background. There were more mouths to feed, however, and the islands, relatively self-sufficient and thus food-abundant in comparison to the mainland, were again used as places of refuge. The postwar years led to reconstruction and economic growth but too few changes in the social and economic systems. The French Civil Code introduced civil organisation and, among other things, provided a route to abolishing the *colonat*. It did not last long, however. Austria immediately reintroduced the system after dispatching Napoleon to a small, notably distant island (having learnt the hard way that Elba was closer to the mainland than it appeared). Moreover, the demands of La Grande Armee put such pressure on the island and coastal economies while they were part of the Illyrian Provinces that the fractured *colonat* and its reestablishment were not remembered with affection.

At the end of the 19th century, island and coastal tenants began to incrementally redeem the land. Sometimes, they were encouraged by special laws, or landlords offered the land for sale due to dwindling revenues. The tenants could pay in installments, and as a rule, the poorer plots on the slopes, which required supporting drywalls, were the first offered for sale. Despite the obstacles, more and more land was redeemed, and owning land became the main aim in the microeconomic policies of the island households. There was also the promise of earning money off the islands, as the migration waves that sent European populations out into other parts of the world for centuries began to affect the islands. Going overseas became a common way of earning money for better land at home. On the other hand, the islands came under the influence of state macroeconomic policies.

Overall governance and economic policies were first created in Vienna (and later in Budapest), then increasingly in Zagreb, and affected the entire Croatian part of the monarchy, including the islands. Changes that had not yet occurred during the war and postwar years gradually came to pass with each new administration. The island specificities, which had been respected by the Venetians, Croatian dukes and kings, and Dubrovnik rulers, were beyond the scope of the economic policymakers in the vast monarchy. For the first time, the islands truly became the objects of mainland-derived macroeconomic and overall development policies, but it soon proved that such policies did not take island specificities into account at all.

Policy measures were applied to the islands as though they were part of the mainland. Their distance, inaccessibility, high unit costs, infrastructure and so on began to be seen as technological or technical problems, topped by the handicap of being islands, which subsumed everything else. The view was that if the island problem could be resolved, if they could be reached as easily as one place could be reached from another on the mainland, then their economic and social issues would fall into place. This type of conceptualised, mainland-made island policy persisted throughout the wars and postwar periods of the 20th century and was, to a great extent, dependent on mainland authorities and whether they cared about the islands at all. It did not change much under socialism, which finally abolished the *colonat* system on the islands but not the mainland approach to problems. It was only under the transition period in the current millennium that any significant changes occurred.

The pattern observed on the Croatian islands, regardless of changing social and economic systems, is rather simple: Mainland-made island policies are more effective the larger the island and the closer it is to the mainland. The larger ones are simply considered extensions (sometimes even anomalies) of the mainland. The smaller ones, however close they may be to the mainland, tend to remain beyond the focus of decision-makers. By the same token, the more important the international overtones of mainland-made geopolitical decisions, the harder the consequences for near islands, both large and small.

The Croatian islands felt the full extent of international politics that were led by and benefited the higher, affluent classes during the establishment, stabilisation and weakening of various pacts that proliferated in Europe in the later 19th century. Among them was the Triple Pact between Austria, Germany and Italy, concluded in 1848. It seemed so fragile that it constantly needed reaffirming and renewing with additional agreements, usually about trade. In late 1891, Austro-Hungary and Italy made a 10-year trade agreement that removed many barriers to imports and exports. News of this may not have reached all the islands immediately, but the consequences were swiftly felt all round. The agreement contained a clause that allowed Italian wine-producers to export their wines to Austro-Hungary, practically without customs duties. The "Wine Clause," as it soon came to be called, proved to be effective, particularly when an official interpretation was added the next year, according to which monitoring the quality of wines exported to the monarchy was left to the Italian authorities. Poor-quality, cheap Italian wines immediately flooded the market, and prices fell by about 30%. The island wines, which had been served at the Viennese court and placed well throughout the monarchy, suddenly lost most of their buyers.

Ten years later, the trade agreement was renewed without the Wine Clause, and it seemed that viticulture and wine-making would recover and even flourish. Phylloxera, the pest brought from the United States that devastated European viticulture in the 1880s, had not spread so quickly, so the islanders had been able to grow healthy vines when the French and Italian vineyards were being devastated. The pest loomed on the horizon and news reached the islands in time, but they nevertheless planted and extended even faster, and their wines sold well in Hungary, Italy and as far away as France. But it was in France, not on the Croatian islands, that phylloxera was finally thwarted and vineyards first recovered.

The wheel of fortune spun quickly, revealing the lack of any top-down regional policy that could have been introduced to save the vineyards. It also revealed that an islander has the ability to spot and use a comparative advantage but the inability to adapt when the advantage ceases to exist. Masters of the sustainable use of resources when insularised and apprentices in economics when caught in mainland networks, they could only observe more and more wines appearing from every quarter, while their vineyards were suffering more and more from phylloxera. By 1925, only a few vineyards remained on the larger islands (Čuka, Mirošević, Faričić and Graovac Matassi 2017). Two evils—one political, which upset the market, and the other natural, on the roots of the vines—caused not only a crisis in the island economy, but also a wave of emigration. The islanders could only continue in the vineyard trade in the United States or Australia. Far from their homeland and separated by the ocean, the islanders listened bitterly to news from the monarchy that their islands had been sacrificed without hesitation to create stronger alliances. Some 10 years later, it was at war again, and this time conscripted the remaining islanders to fight for the imperial cause.

Historians of the Austro-Hungarian period cannot deny any of this. They may shrug their shoulders and say that island economies were collateral damage that the empire could have done without. They may also claim that Vienna and Budapest did in fact care about the islands, because a great deal of maritime infrastructure was built in the late 19th century, right up to collapse of the monarchy in 1918. At the same time, they would have to admit that this was not done for the benefit of the islanders, although it did provide some positive externalities. The islands became tools in the extension of monarchical interests, as by entering the Triple Alliance, Austro-Hungary was bound to help its allies in military terms as well. Among other things, it changed its outdated maritime strategy and started strengthening the fleet and securing safe navigation in the Adriatic. From Pula in Istria to the Bay of Kotor in Montenegro, fortresses and harbours were built, fleets assembled and navigation routes taken into consideration (Stulli 1959). The outer islands gained lighthouses and safe harbours, and in many unsheltered coves, stone jetties were built to provide shelter for larger vessels. Navigation throughout the temperamental Adriatic Sea became safer. Historians of the Austro-Hungarian Monarchy and the later Kingdom of Yugoslavia might also add that the islanders who left for Australia and both Americas at that time sailed out of the Adriatic safely.

AT WAR

The Kingdom of the Serbs, Croats and Slovenes, later the Kingdom of Yugoslavia, inherited islands from Austro-Hungary equipped with infrastructure, although without a properly formed island development policy. The islands experienced only the politics of the great powers: In late 1920, Istria, the Kvarner islands, Cres, Lošinj and the Lošinj subarchipelago in the northern Adriatic, and Lastovo and Palagruža in the southern Adriatic were ceded to Italy. Two mainland-made island policies came about in the period between two world wars. The first was an Italian policy that was completely irredentist and the second a Yugoslav policy that was not oriented toward maritime management or coastal development, let alone island development.

The islands that belonged to Italy remember this as a time when resources were opened up to Italian investors and unremittingly exhausted, and a general Italianisation was imposed in culture and society. The islands that belonged to Yugoslavia recall scarce links with the mainland, less-than-optimal maintenance of the Austrian maritime infrastructure, political pressures from Belgrade and a general regional policy focused on development of the continental areas. The island collective memory recalls vineyards devastated by phylloxera and unmitigated emigration. Tourism began to develop without a directing policy, while shipbuilding and seafaring were treated with indifference. The economic policies of both the Yugoslav and Italian kingdoms did not record any specific measures relating to the islands. This was only to be expected of the Banat of Croatia (*Banovina Hrvatska*), which was established in 1939, as a Yugoslav region with sufficient autonomy in economic policy; however, not quite two years later, the Second World War descended on Yugoslavia.

The Second World War affected the Croatian regions more than the First World War. Yet, the islands did not suffer significant devastation. After the Kingdom of Yugoslavia collapsed in 1941, most of the islands belonged to Italy, which took them without opposition, while Brač, Hvar and part of Pag found themselves in the Independent State of Croatia (NDH). The terrors of living under occupation and the burdens of the war economy were felt by everyone, and concentration camps were established on Rab and Molat. On some islands, sporadic resistance was organised, but it was soon shown that the Adriatic islands, like other islands of their size, were simply too small to accommodate guerrilla warfare. The islanders joined the war on the mainland and discovered that after the Italian capitulation in 1943, the occupiers had left the islands peacefully and surrendered them to the People's Liberation Army (PLA) of Yugoslavia. The German offensive soon followed and, on its heels, incursions by the army of the Independent State of Croatia, so the PLA withdrew from all the islands except Vis. In the spring of 1945, all the islands were recaptured by the PLA. On most, there was combat that ended with the German or NDH units being driven out or taken prisoner.

For the first time in history, the Adriatic islands were not only refuges, but also part of the battlefield. The outliers retained their geopolitical significance. After the PLA took Vis, following the Italian capitulation, British troops landed. They built an airport and set up fixed communications with Allied forces in Italy. The German army occupied mostly undefended larger islands but made no attempts regarding Vis. The bloody lessons of the Second World War were learned in the Adriatic, too: Guerrilla resistance might have been nearly impossible on the occupied islands, but an island is quite hard to take if properly fortified and prepared to defend itself. Vis, at the entrance to the Adriatic, remained an Allied military base until the end of the war, a mustering point for the departure of civilians from its own settlements and other islands to free parts of Italy and Egypt. El Shatt refugee camp on Sinai is another vivid memory in the island collective as an oasis, or what Herodotus called an island in a sea of sand. It appears in many islanders' personal documents as their place of birth. In 1946, when the camp was disbanded and the refugees sent home, they returned to real island life, surrounded by the sea.

After the war, Vis and nearby Lastovo, which was equally important in strategic terms, had to face the reverse side of the medal they were decorated with. The People's Liberation Army, which grew to become the Yugoslav People's Army, thought the islands should remain army bases and fitted them with a network of defence objects. Following military logic, they also prohibited access to foreigners. Tourism in Yugoslavia, which started to develop in the late 1950s and continued until the end of the 1980s, bypassed both islands. Foreign access was only granted in 1989, and the Yugoslav People's Army withdrew in 1992, several months after it had abandoned the other islands. Vis and Lastovo, delayed 30 years, began to develop their own tourism. The Adriatic is rather small (the Venetians called it a bay), but its islands, some near, some not so near, have possessed strategic importance for centuries. Braudel wrote that La Serenissima thought of them as her fleet (Braudel 1972). The British navy held on to Vis throughout the Napoleonic Wars and in the Second World War, and the Yugoslav People's Army liked to stress that the islands were their aircraft carriers and impenetrable obstacles to any invaders who might want to set foot on these shores.

SOCIALISM

During socialist Yugoslavia, concern for the island specificities remained more or less the same. This time, the postwar period was marked by an upturn in the economy but in the socialist mode. The economic policies conducted in the Soviet Union after the First World War were imposed on the new socialist countries after the Second World War. They were followed without hesitation. The policy of "primary accumulation of capital" was implemented throughout the country, regardless of regional differences, and the main aim was to develop heavy industry as rapidly as possible. Centralisation and control of enterprises were crucial, while the market, as a result of the widely accepted critique of capitalism, was reduced as far as possible and made the focus of centralised planning. In a country that emerged from the war with 70% of its population engaged in agriculture, these people were considered a massive workforce pool for struggling industries in need of more workers than cities and towns could supply. Peasants were encouraged to leave their villages for the cities and become manufacturing workers. Everything possible was done to boost the construction of basic infrastructure, specifically the railways and development of the heavy industry sector.

Those who stayed in villages were forced to accept lower than market prices and sell their produce to state chains of stores that sold them on to industrial workers for marginally higher prices. The price of labour was thus low, and a greater proportion of revenues could be earmarked for investment. Investments were made in industry, and the new factories needed even more workers who could make both ends meet with low wages, so there was room for even more investment. Judging from the growth rate, which in the early 1950s sometimes reached double figures, everything was going according to plan. In the flush of postwar renewal, the attention of socialist planners and the spotlights of socialist propaganda were turned on factory chimneys, labour victories, growing industrial centres and urbanisation. The inhabitants of once-overcrowded villages went to overcrowd the cities, abandoning small-scale agriculture that had been partially transformed into massive state-owned enterprises. The villages, most of them no longer fully populated, remained in the shadows.

The islands also remained in the shadows and were regarded primarily as reservoirs of potential workers but not places where the socialist economy could take root. Larger islands attracted attention from socialist planners, and some became the sites of coastal factory outposts. Pag, Šolta, Brač and Korčula were too far from the mainland for daily commuting but near enough for such outposts to launch production and, if necessary, hide losses. In a way that may seem naïve today, socialist politics attempted to ensure the even spatial distribution of the working class throughout the country, including the larger islands; however, this attempt at social engineering did not include the smaller islands. Some were literally incorporated into the mainland to increase development space.

The best example is the island of Krk, whose inhabitants watched as the coastal city of Rijeka, situated on a narrow belt of land between the sea and the steep mountains at its back, spread along the shoreline until it reached the strait dividing their island from the mainland. The only way Rijeka could expand further was via Krk. In the 1970s, planners proposed and built a bridge across the strait, completing it in 1980 to great fanfare (Institute of Economics, Zagreb 2018). Several industrial plants were built on the north part of the island close to the bridge. Krk demonstrated that an island is as close as the mainland needs it to be. In part, at least, it has become mainland. Thirty years after the bridge was opened, a cluster analysis was produced of 36 municipalities and towns in Primorje-Gorski Kotar County, which covers part of the northern Adriatic coastline, the city of Rijeka, the mountainous hinterland and the Kvarner islands (Institute of Economics, Zagreb 2018). It showed that the municipality of Omišalj, nearest to the mainland on Krk, had the features of a mainland municipality. In future reports on the implementation of county spatial plans, Omišalj will be regarded as a mainland municipality.

Despite the proximity of the mainland, the Croatian islands have not proved to be suitable places for the development of industries that might play any important role on the mainland. The exceptions are the saltworks on the island of Pag and in Ston, a settlement on the Pelješac peninsula, which together meet about 35 to 40% of Croatian salt consumption, and high-quality construction stone quarried on Brač for export (Starc et al. 1997). Canneries used to be another exception, but their days are gone. Of 59 canneries on the eastern side of the Adriatic, 32 operated on islands and exported canned fish throughout Europe. Freezer technology put an end to this industry. Yields are still favourable, but the fish end up in cold storage vans that await them on jetties in fishing harbours, only to leave for the inland. Only three canneries, two of them on islands, are still operational today (Starc et al. 1997). So, industrial resources have been reduced to nonmetal minerals. Almost all the islands had quarries and saltworks sufficient for local needs, but few have been able to increase production above the local demand.

Factory outposts on the islands that used mainland resources were doomed to fail. All the production input had to be transported to the island, and the output returned to the mainland, increasing transportation costs. Increasing the workforce throughout the islands was simply unaffordable. On the other hand, it became possible to commute from the islands nearest to the large coastal towns for work. The people of Krk started going to work in Rijeka or Kraljevica, crossing paths with people from Rijeka or Kraljevica going to work on Krk. The people of Ugljan and Pag went to Zadar, the people of Korčula to Pelješac and the people of Šolta to Split.

The only economic activitiy that could develop and bear the high costs of transportation was tourism. Until the 1980s, when connections with the mainland increased, island tourism developed more slowly than in the popular coastal resorts. It was harder to build and maintain than on the mainland, but the islanders managed to respond to the steadily growing demand for sun, sea and unspoiled nature. Island tourism, which was mostly locally planned, quickly outstripped the industrial sectors planned for the islands in the state planning institutions on the mainland. This was only true of the larger islands, however—the smaller ones lagged behind again and could only provide accommodations for visitors whose expectations were not too high and who did not mind constant reminders to save water.

Ironically, the subarchipelago of the Brijuni islands, once the winter residence of the Austrian imperial family, who spent their summers in Vienna, was not developed at all. In the late 1940s, it was claimed by the party leadership and turned into the official country residence of President Tito. Not intentionally, perhaps, they showed that islands could provide the highest level of refuge even under socialism. The subarchipelago with the best tourism infrastructure at the time was forbidden to locals, let alone foreigners.

TOWARD A TOP-DOWN POLICY

Socialist Yugoslavia had a four-tier political and administrative system. The federation comprised republics that were divided into associations of municipalities and then municipalities. The system began to decentralise in the mid-1970s, and central planning lost some of its grip, to be abandoned by the 1960s. The islands (except those under military supervision) were left to the federal unit to which they belonged, that is, the Socialist Republic of Croatia (SRC). As the federal units also gradually decentralised, managing development was increasingly left to the municipalities. The nine island and six coastal-island municipalities of that time had a free hand in adopting their own development programmes and spatial plans. To some extent, in the 1970s and 1980s, islands conducted local development policies with obvious respect for their specificities. At the same time, the overall development policies of the SRC were still being produced to cover every land area, including the islands.

Many complaints could have been made about negligence concerning island specificities, but one question was never ignored: how to maintain the integrity of the territory and waters of the SRC, that is, how to connect the islands with the mainland. In contrast to the mainland traffic policies of the other Yugoslav republics, Croatia always had to include a maritime component. This aimed to provide the means for daily travel from each inhabited island to the mainland and back. It was clear that changes in socioeconomic systems and ideologies did not necessarily lead to changes in attitudes toward the islands. In the SRC, the island "handicap" was made explicit and uncritically emphasised as something that needed to be "overcome" to resolve other development issues. The idea of the islands as a development potential, rather than an obstacle, only began to appear in the late 1980s.

Postwar socialism is easier to describe and understand than the version that dominated Yugoslavia and Croatia beginning in the 1960s. A simplistic, crude system with clear goals evolved into what was known as "self-management socialism," with many developmental protagonists required to manage the simultaneous coordination of vertical and horizontal development at every level. Paradoxically, this was supposed to happen within a one-party political system where the League of Communists always had the last word. This breed of socialism differed from the Soviet one and was constantly criticised from the east. Meanwhile, tourists from other Eastern European countries showered praise on it as they enjoyed summer holidays on the Adriatic. The party leaders called it "socialism with a human face." A tragic reversal took place when Yugoslavia collapsed and the conflicts that followed affected the mainland much more than the islands. This time, the islands were not part of the battlefield and were not devastated, but they once again became places of refuge, halted their tourism activities and opened their hotels to displaced people. Once again the islands had to wait to be affirmed on the field of development policies. This was ironic, as less than five years before the war they were finally made the subjects of a mainland-made island policy that tried to acknowledge their specificities. In the late 1980s, appropriate institutions were established at the level of the republic.

This began in 1985, when the Association of Municipalities of Dalmatia launched an initiative and proposed a *Social Agreement on Development Policies for the Adriatic Islands*. It was signed by the Dalmatian islands and coastal-island municipalities, the relevant public utility infrastructure bodies and the government of the SRC. The Association of Municipalities of Rijeka, which covered the northern islands, did not launch any initiative at the time.

In the same year, an Inter-Island Coordination Commission was formed, comprising 15 presidents of the executive councils of island and coastal-island municipalities, two presidents of the coastal-island municipalities of Rijeka and Dalmatia, and the president of the Executive Council of the Government of the Socialist Republic of Croatia. The reader needs to be reminded here that the Yugoslav political systems and self-management economy were unique, and that specific institutions were established. The terminology and the political jargon were also specific. A more thorough explanation of the role of social agreements, executive councils, social plans, coordination commissions, and so on would require at least another chapter in this book. At any rate, the institute of the social agreement was abolished in 1990, when Croatia became independent. The island social agreements therefore ceased to be valid, and the Inter-Island Coordination Commission was disbanded.

The associations of municipalities formed in the mid-1970s established institutes for social planning that produced annual and periodic social plans. These plans were the basis on which the island municipalities adopted their own social plans, which were intended to direct economic and social development. Thus, at least in formal terms, the system of socioeconomic planning was completed. Municipal spatial plans were also supposed to be produced to complete the spatial planning system. There was, however, no development strategy for managing overall island development prepared and adopted at the level of the SRC. Actually, those responsible for the island policy in the nascent state did not need one. The budget item was earmarked, and the body in charge assumed the discretionary right to distrubute the money. In the socialist self-management and yet one-party system, it was deemed sufficient.

The need for strategic documents at every administrative level was only observed in spatial planning. The first Spatial Plan of the SRC was adopted in the mid-1970s, and the second in the late 1980s. Both plans depended methodologically on previous spatial plans for the Adriatic. In the period between 1967 and 1972, a tripartite Plan and Programme for Spatial Planning in the Adriatic Area was adopted covering the northern, central and southern Adriatic. In these plans, for the first time, planning the maritime area as an area of special interest was introduced.

The spatial plans of the municipal associations had to be aligned with the Spatial Plan of the Socialist Republic of Croatia, and municipalities had to align their spatial plans with those of the municpal association. In this system, the mainland-made island policy between 1985 and 1990 was reduced to decisions on infrastructure projects that the Inter-Island Coordination Commission made at the level of the archipelago according to rather vague criteria. Investment data could no longer be gathered, as projects were financed from the state budget and the budgets of public infrastructure companies who invested separately in the island water supply, sewage, electrification, additional ferry lines and the like. In the melting pot of new independence, changes of government and the establishment of new institutions, socialist planning documents, along with implementation reports, ended up in the waste bin. Only old volumes of the *Official Gazette* remain, with reports on the implementation of the budget of the SRC for 1985 to 1990, which indicate that approximately 50 inhabited islands were to receive about 30 million Deutschmarks.

THE ISLANDS OF THE REPUBLIC OF CROATIA

Mainland-made island policy continued after Croatia gained independence. The new incoming leadership did not intend to proceed with any socialist practices, and little attention was paid to policy documents found in the offices of the disbanded socialist institutions. On the other hand, socialist officials were not dismissed. Most of them kept their jobs in the state administration or were reassigned to newly formed ministries. Those who had worked on island policies also kept their jobs, as did some external consultants. Experience was lacking all around, but the change of government and the war did not reduce their intrinsic motivation to deal with the islands. Unexpectedly perhaps, intrinsic motivation appeared as one of many components of islandness, indicating that islandness can be developed on the mainland. It has been demonstrated that many people on the mainland had a good understanding of islands and strong motivation in building island policies (it has also been demonstrated that there were many islanders who had no such motivation at all). What is more, in the euphoria of independence and the strivings of the new state, it seems that the establishement of new and better island development policies would be only a matter of time.

Two initial tasks that soon proved crucial were carried out. The first was to set up a body within the state administration that would be responsible for managing island development. It proceeded surprisingly smoothly: It was much easier in those days to get an appointment with a minister, so the minister of maritime affairs agreed from the outset that his ministry should set aside resources, officials and office space. During the mandate of the third Government of the Republic of Croatia, the Centre for Island Development was thus established at the Ministry of Maritime Affairs. The second task seemed more difficult but actually proved easier than the first. A new constitution was being feverishly developed, and many had an opportunity to propose amendments. It was enough to have a personal contact with one of the president's advisers to arrange a meeting about the islands. The adviser showed interest and, at one brief meeting, listened to the proposal, picked up the draft constitution, added the islands and hurried off. One of the authors of this chapter was standing next to the adviser at the time, still breathless from trying to convince him, and watched in enchantment as the islands entered the constitution. The importance of this off-the-cuff meeting was confirmed a few months, and even years, later. The constitution often had to be pushed under the noses of all those who were against the adoption of the Islands Act.

The draft was revised and amended from day to day, but the island issue was never questioned. The Constitution of the Republic of Croatia, known as the Christmas Constitution, was adopted on 22 December 1990. Article 52 states,

> The sea, seashore, islands, . . . and other natural resources, . . . which are specified by law to be of interest to the Republic of Croatia shall enjoy its special protection. The manner in which any resources of interest to the Republic of Croatia may be used and exploited . . . shall be regulated by law.

THE NATIONAL ISLAND DEVELOPMENT PROGRAMME

It would be a cliché to say, "The rest is history," but the basis for island development policy was now established. Further steps required more effort, but communication with other ministries and public companies was much easier. The next step was made in 1994, toward the end of the Homeland War, when the Centre for Island Development was moved to the newly established Ministry of Reconstruction and Development. The new ministry had a significantly larger budget and estabalished the Directorate for National Development Projects, which launched and financed the production of national development documents. There were enough supporters of island development to convince the deputy minister who led the directorate that the island development document was urgent enough.

In the spring of 1995, the Council for Island Development was formed, comprising 25 members, of whom about a dozen actually set to work. The first problem they discovered was "hidden data," because the larger islands had one or more municipalities, while the smaller ones were parts of municipalities on the mainland or another island (see Chapters 3 and 6). Official statistics are gathered at the level of municipalities and do not account for further disaggregration, so that data on the small islands was "hidden." The council discerned a chronic problem (the Croatian Bureau of Statistics has yet to resolve it) and agreed that any analysis would, of necessity, be incomplete. Nonetheless, the task was carried out. In late 1996, after many visits to islands and institutions that were supposed to care for them in one way or another, the council prepared the National Island Development Programme (NIDP). Parliament adopted it in February 1997.

The programme was a comprehensive, inspiring development document that showed the current state of affairs on the islands and listed their comparative advantages, while assessed their limiting factors. Like many development documents of the time, it concluded,

> The islands have arrived at a developmental crossroads. They can take the path of sustainable development and finally become a special, attractive place for

> investment and life in general, or they can take the shortcut of quick and unreasonable exploitation, which will irretrievably impoverish and relegate them to a marginal status in Croatia and the Mediterranean. (Starc et al. 1997, 71)

For the first time in the Adriatic, in a single island development document, the principles according to which island development should be directed were defined. From these principles, developmental goals were derived. The following six principles were outlined:

1. *Islands are equal parts of Croatia in terms of development policy.* This principle finally overcomes the negligence in development policies that has accounted for the division into islands and the mainland, and "large" and "small" islands.
2. *An island is a system consisting of the island ecosystem, economy (all activities of the people relating to production, consumption and traffic) and the community of people living there.* An island appears as a system that can only be successfully managed as a whole.
3. *Island resources must be sustainable and fully used.* The investments encouraged on the islands will be those that are ecologically, economically, technologically and socially sustainable. Any use of island potential that is partial or unsustainable in some form threatens to unbalance the island system and leads to less-than-optimum use, and even to fatal, total exhaustion in the long run.
4. *An island is a development unit whose development is the primary concern of islanders.* The state and coastal-island counties provide only for necessary institutional conditions and incorporation in the national infrastructure and superstructure systems.
5. *State and county administration, and local self-government, should take an active role in island development.* The state must be present on the islands that require revitalisation and must actively invest in infrastructure and superstructure as a special entrepreneur who, from the position of integral island development, encourages and directs other entrepreneurs to invest in sustainable development and prevents investment in unsustainable development.
6. *There must be continuity of island development management.* This principle commits to continuous care for actions initiated, which means that the initiators of island development policy must fully implement the programmes they have started and select, use and modify, in the long run, development instruments that will direct users of island resources toward their full and sustainable use. (Starc et al. 1997, 78–80)

From these principles, the goals of managing island development were derived, followed by specific tasks that needed to be accepted by particular state and county institutions. Eight basic goals were listed:

1. *Ensuring sustainable development.* This is the principal goal and purpose of island economy and ecosystem management. The basic goal is thus determined as the steady and fully sustainable use of island potential at an economic, ecological, technological and social level.
2. *Maintaining socially adequate population density.* An island is adequately populated when the population number, age, sex and qualification structure, as well as the way of life and standard of living, create a community that is appealing enough to not only retain its members permanently, but also attract new inhabitants, reproducing without the direct measures of state development policies. Such communities existed in the past, although with a different standard, on almost all the Croatian islands.
3. *Retaining the present population, encouraging families to have more children and returning of younger, economically active people who have moved to the mainland.* These are the demographic goals of island development, fully coordinated with the goal of socially self-sufficient population density. Since the small islands' populations are dwindling, these goals can be accomplished by establishing conditions to resettle people from other places, if the islanders cannot bear the burden of development by themselves.
4. *Balancing island investment conditions with the conditions under which investments take place on the Croatian mainland.* This refers to the economy, public standards and everyday island life in general, which does not make investments in the islands appealing today. This goal can be realised primarily by infrastructure construction and will be attained when the current lack of island links to the mainland and links between and within islands no longer scare off investors or prospective new islanders from settling.
5. *Diversifying the structure of the island economy.* This is a further goal of island development. The more activities an island has and the more production units within these activities, the easier it will be for the island economy to overcome the ups and downs of island production demand.
6. *Maintaining versatility of island households, the basic units of the island economy.* The more activities its members are involved in, the more stable households are at the economic and microsocial levels. The same applies to the island companies and cooperatives.
7. *Ensuring full physical accessibility to island resources.* Accessibility is a prerequisite for resource management, as development measures can only

be coherently formulated and carried out successfully if every part of the islands is easily accessible. Physical accessibility will be achieved when all the island agricultural land and other areas used for production or consumption purposes have adequate road networks.

8. *Ensuring accessibility of island values in legal transactions*. This will be achieved when property is properly registered and boundaries are established, and when property or at least the right of use are indisputable, that is, when real estate transactions in the entire territory of an island are legally feasible. (Starc et al. 1997, 78–80)

The drafters of the NIDP saw immediately that the principles and goals were easier to compile than the actual content of the programme. The issue at hand was not so much the islands, but the island space, and it was necessary to determine how far the programme should extend to the sea and perhaps also to the mainland. The Adriatic near islands are similar to peninsulas in many ways, and it is not always easy to demarcate borders. So the programme was extended to the mainland:

> The 62-km-long Pelješac peninsula is usually listed together with the islands. With its 348 km^2 it lies in the same direction as the surrounding South Dalmatian islands. Pelješac has the same structure as the islands, the same climate, and its flora and fauna are more like those of the islands than the hinterland. (Starc et al. 1997, 8)

A response to the issue of the island waters was given in principle:

> Island management and its goals are not necessarily exhausted on the island proper. Island management extends to the sea and seabed around the islands, the traditional space of island life and economic activity, where islanders should do no less than they have always customarily been able or permitted to do. (Starc et al. 1997, 80)

Although many objections may be raised to this definition of island space, it was enough to produce the principles, goals and then tasks and assignments for the state, county and local units of self-management. The tasks and assignments, with specified executors and financial sources, were divided into five groups:

1. Adopting the Islands Act
2. Listing activities of the Centre for Island Development at the Ministry of Reconstruction and Development
3. Preparing and adopting sustainable island development programmes, physical plans and targeted scientific studies

4. Adopting and amending existing laws relevant to island development
5. Preparing and adopting national infrastructure and superstructure programmes for the islands

THE ISLANDS ACT

Frequent meetings with ministries and public companies, mostly located in Zagreb, convinced the drafters of the programme that a development document would not be sufficient to launch island development policies. In 1997, Croatia had already produced and adopted about 30 development documents with such names as "National Strategy," "Strategy of the Republic of Croatia," "National Programme" and so on. When the Islands Act was adopted in 1999, there were twice that number. In 2018, when the new Islands Act was adopted, there were about 260. A preliminary analysis showed that most did not contain action plans or include confirmed sources of financing, nor did they define how the effects of any measures were to be monitored. There is no evidence that any of these documents has actually been implemented.

In 1997, no one would dare to predict the progress of development documents, but it was clear that the most urgent task was to adopt an act that would raise island development to the highest possible degree of obligation; however, those responsible in the ministry thought it could wait. The stalemate lasted for a couple of months, and the drafters of the NIDP decided to turn to the opposition. It was clear that the main opposition party would readily propose an act that should have been proposed by the party in government (how much they really cared about islands was shown during their 2000–2013 term in government).

It is always easier to contact members of the opposition in Parliament than those in power, so a meeting with an opposition member of Parliament (MP) was quickly arranged. Knowing that only three MPs were required to propose a bill, the member readily looked through the NIDP and accepted the challenge. Less than a month after the meeting, the Islands Act was proposed to Parliament.

The news spread quickly and reached those who were supposed to hear it, and in the ministry where the programme had been produced, a decision was made to not wait any longer. The stalemate ended, phones started ringing and a couple of drafters of the NIDP, still active members of the Council for Island Development, were given a new task. They looked through the programme and produced another draft Islands Act. In the autumn of 1998, something unusual happened in Parliament: Two proposed drafts of the same law were on the table. The opposition was easily outvoted, and the Islands

Act was adopted in April 1999 (OG 34/1999). It was proudly noted that in this respect, Croatia was third in the world, behind Japan and Finland.

The adoption of the act meant that the first of five groups of tasks and assignments in the NIDP had been fulfilled, and the act itself was supposed to ensure the implementation of the other four. The Ministry of Reconstruction and Development was designated to implement the Act and the Centre for Island Development was the competent body, with its tasks already defined in the programme. This completed the second group of tasks and assignments, and the centre moved on to the remaining three. This included organising the division of the islands mentioned in the act. The usual geographic division, which did not take into account development level, had not proved appropriate, so the act divided the islands into only two groups:

1. Forty-four underdeveloped, inhabited or occasionally inhabited islands
2. Inhabited islands not included in the first group and Pelješac peninsula

The first group included all the small islands, with Vis and Lastovo, which were deemed less developed than other large islands due to their isolation, imposed by the former Yugoslav military. The debate on what size an island had to be to be considered small had been going on during the preparation of the programme and was never satisfactorily concluded. The only agreement reached was that, in the Adriatic, islands with surface areas less than 30 km^2 and fewer than 300 inhabitants should be considered small.

As a result of the programme, the act stipulated the production of 26 Sustainable Island Development Programmes (SIDPs). Each would deal with the overall development of the islands, covering all activities on them, and proposing activities and projects in which they might become engaged. The centre was to launch and finance the preparation of SIDPs. In the meantime, it had grown into the Islands Directorate, with a deputy minister at its head. That year, the Regulation on the Contents and Methodology for Producing the Sustainable Island Development Programme was adopted (OG 94/2002), which prescribed the participation of all island development stakeholders and the adoption of specific measures, accompanied by proposed methods of financing.

The act also stipulated the preparation and implementation of 14 State Programmes for Island Development (SPID), which would deal with individual sectors and social activities on the islands. The state programmes dealing with traffic links between islands and the mainland, and the islands themselves; the water supply and treatment of waste water; electricity supply; island telecommunications; health care and telemedicine; preschool, primary and secondary education; and the organisation of the cadastre and land registry were considered the most important. The Ministry of Environmental Protection and Spatial Planning was to have the greatest workload. Apart from being the competent body for spatial planning and the supervision of construction on the islands, it was to propose one state programme and participate in the implementation of five of them.

The act also prescribed the Annual Island Programme, which Parliament was to adopt each year. The Annual Island Programme was to include all projects from the SIDP and the state programmes to be financed in the coming year from the state budget. The Annual Island Programme was to remove any potential overlap and align the state contributions for the islands, and Parliament had an instrument for guiding island development policy with both bottom-up (SIDP) and top-down (SPID) components. The Islands Directorate was to prepare the Annual Island Programme, with the proviso that Parliament had already saved them some of the work. This was because the three island national parks, Brijuni, Kornati and Mljet, were not included in the Islands Act, as they were administered by the Ministry of the Environment, and the park management and spatial plans were adopted by Parliament without the participation of the island municipalities and towns within their boundaries.

The administration of the parks, and particularly the drafting of and attempting to pass the SIDP and state programmes, revealed the complexity of conducting island development policies. During the drafting of the Islands Act, it was already clear that the Islands Directorate needed to push for and monitor the implementation of the NIDP and the act; however, due to the large number of bodies involved on both the islands and mainland, it was clear that the directorate could not and should not be a single bearer of island development policy at the level of the state. The main task of the Directorate (it only had a task or two to implement itself) was to link and coordinate the other bodies, monitor them and evaluate their work. One of the rare tasks the directorate had to complete on its own was the drafting of the SIDPs, and this was suitably accomplished. The programmes were produced between 2003 and 2005, some quickly, some slowly, and several were ready by the end of 2003. The act stipulated that they had to be adopted by the island municipalities and the government.

In late 2003, the government adopted only one SIDP, for the island of Šolta. The next government did not adopt any, and the island municipalities did not respond. The state programmes foundered in the drafting stage. Since they were sectoral programmes, 16 ministries and public companies were involved, but the Islands Directorate could only encourage them and remind them to implement the act. Eventually, only two state programmes were adopted. During the first years of the act, the directorate kept postponing the

Annual Island Programme to the next year. It soon gave up even doing this, as it was obvious that in spite of demands and warnings, it would not be able to gather the input data.

Although it was hailed as the first act to deal with the islands and was one of only a few based on a previously produced programme, the Islands Act has been difficult to implement. In the first place, it has proved to be too complicated for the existing institutional capacities of state administration, although the preparation of the SIDP showed that it was possible to gather enough experts together to produce the necessary development documents in line with modern development methodology. These experts were also found in public companies, so that the lack of state programmes could not be ascribed to methodological incompetence, but to a lack of willingness to cooperate between ministries and public companies. On the other hand, the Islands Directorate had the legal authority to coordinate and assess the other stakeholders, but it could not force them to do their part of the work. The unification of 26 SIDPs and 14 state programmes in a coherent mainland-made island development policy failed.

Life continued on the islands, however, and in the Islands Directorate and other state institutions throughout Croatia. Many ministries and public companies continued to work as they had before the act, maintaining and building infrastructure, transporting the islanders to the mainland and back, providing health care and education and so on, according to their understanding of the needs and in line with their plans. The act made to align all this was only implemented sporadically, the opposition was no longer interested and no one asked about the effects of the budget line assigned to the Islands Directorate. It reached about HRK 200,000,000 (EUR 27,000,000) in 2007, and the Islands Directorate assumed the discretionary right to allocate it throughout the archipelago. To provide some sort of legitimacy for this allocation, the island mayors were asked from time to time, but at least once a year, what their islands needed and what the priorities were. They answered promptly, and some even found dusty copies of the SIDP in drawers and looked for things to suggest. The directorate received projects and estimates of costs, which, when taken together, always exceeded the available budget. So the funds would be allocated according to criteria that in political jargon are called "nontransparent," and those who were not given funds in one year would adapt their project demands and estimates, waiting for the next round of questions and answers. For the second time, the islands found that those on the mainland who were making decisions about directing and financing their development did not actually need their programmes. Budget items were earmarked, and there was a body that decided on their allocation. In the young, supposedly capitalist, multiparty system, it was deemed sufficient.

If the reports on the implementation of the Islands Act submitted each year to the government are taken together, the impression is gained that the state has been investing significant sums in island development and that beneficial effects are being achieved. The 2017 report lists all public expenditures for the year, along with 13 ministries, 7 coastal-island counties, 12 harbourmaster's offices, 7 larger coastal-island local self-government units (LSUs), 1 of the 3 island national parks, both island nature parks and 15 public companies (Ministry of Regional Development and EU Funds 2018). It also mentions various European Union (EU) funds and development loans given by the Croatian Bank for Reconstruction and Development and the Council of Europe Development Bank. Between 2006 and 2017, the average annual nonreturnable funds invested from the state budget and EU funds in island development (75%) and average credit investment in island development (25%) totalled HRK 1.6 billion (some 216 million EUR) (Ministry of Regional Development and EU Funds 2018).

The significance of these investments remains dubious, because investments in other parts of Croatia that also receive state aid are not mentioned in the report. The total level of such allocations has never been reported either, so the islands have no way of telling whether they were receiving more or less than other aided regions. Of course, this information exists, but it is scattered among ministries and public companies, and the Islands Directorate is unable to collate it due to poor horizontal communications. Furthermore, this information can only serve as an input indicator to show what has been invested where and when. There are no output indicators in the reports, so it remains unclear whether investments have produced the required development effects and whether they were sufficient, or perhaps unjustifiably high (the islanders would, of course, deny the last). It is also unclear whether the development effects (the islanders would have to admit that there have been many) can be attributed to investments from the state budget (Islands Directorate) or some of the many others on the long list of state development bodies. The island LSUs are also involved in development and can request loans and EU funds without needing to inform the state administration. On the other hand, the reports do not mention criteria for allocating funds and launching projects, so it remains unclear why some development projects on certain islands have been financed and others not.

It is no wonder that the islanders mostly tacitly accept "nontransparent" development policies. Their administrative division reflects the historical orientation of particular subarchipelagos toward the mainland and poor communications with other islands. The Dalmatian island of Čiovo stands out in this regard. It is separated from the mainland town of Trogir by a strait measuring some 100 metres, across which two bridges have been built. The island has a surface area of 29 km² and approximately 6,000 inhabitants, but in administrative terms, it is tripartite: One-third belongs to the town of Trogir, one-third to the city of Split and one-third to a municipality on the island. The island, which according to the NIDP "appears as a system which can only be successfully managed as a whole," is subject to 3 spatial plans, 2 town plans, and 1 municipal development plan, and the island provisions of the development strategy for Split-Dalmatia County. Since Split and Trogir belong to the urban agglomeration of Split, the Strategy for the Development of the Urban Agglomeration of Split is implemented on two-thirds of the island. There is also a SIDP that (perhaps fortunately) is not being implemented. This kind of administrative division of the Croatian islands requires some skilled surfing by local self-government units, and experience shows that the LSUs on the more developed northern islands are more adept. There are, however, LSUs on southern islands that cannot finance their own operations and depend to a greater or smaller extent on the state budget. The situation on the mainland is no better. Croatia has 556 local self-government units. About 40% of them are unable to cover their own expenses. It is obviously not advisable to criticise a system that can but need not finance their development projects.

THE NEW ACT

The Islands Act of 1999 was implemented as far as officials were able and wanted to, and as far as politicians willed. The drafters of the act can be criticised for not analysing the institutional capacities of the implementer and overburdening them with too many tasks. A regulatory impact assessment would have helped to show who would be impacted by the new act, but at the time, this was not undertaken. Regulatory impact assessments only became obligatory in 2013, when Croatia acceded to the EU, and only became part of the procedure for proposing the act several years later.

The implementation of the act therefore lagged behind in terms of vertical and horizontal communication in cooperation with a large number of institutions at all three administrative levels of the Republic of Croatia. In terms of normal administrative tasks and procedures, the act was implemented and quickly incorporated as legislation that made everyday island life easier. Coordination was, however, poor and communication lacking. The provisions on transportation connections were implemented, but they were directed by the Ministry of Maritime Affairs, which occasionally informed the Islands Directorate. Provisions on the prohibition of importing foreign species to the islands and small-scale fishing without compensation were directed by the Ministry of Agriculture, which did not inform the directorate. In addition, state bodies left real estate owned by the state to units of local self-government in line with the NIDP and the act but in an uncoordinated way, without the knowledge of the directorate. Uninhabited islands, where the state had the right of primary purchase according to one of the amendments to the act, were recorded in the cadastre and proprietorship documents without the knowledge of the directorate. On the other hand, the Croatian Bank for Reconstruction and Development issued favourable loans to entrepreneurs who were launching or extending activities to make island development sustainable (the act defines the relevant activities) and cooperated with the directorate.

The directorate implemented provisions on compensation paid to islanders for island road transport and ensured that entrepreneurs on undeveloped islands received more funding than those on developed islands, and compensated water carriers for expenses related to transporting water to the small islands. Implementation reports on the act, which were delivered annually to Parliament, were therefore as full as the participation of other bodies allowed. The directorate also prepared the procedure for acquiring the Croatian Island Product label, granted to island entrepreneurs who produce authentic island products. Since 2015, the directorate has issued public tenders for the allocation of budget resources that are available to island entrepreneurs and LSUs, and the commission comprising members from every administrative level decides on allocations. Along with the well-received Croatian Island Product label, this step toward the transparent financing of island projects is perhaps the most significant achievement of the Islands Directorate.

Implementation, in which the directorate does as much as it can but much less than what is necessary, and in which many significant provisions are carried out by others (the key provisions have not been implemented by anyone), can therefore be ascribed in part to the restructuring of the state administration, which happens every four years, immediately after elections. The Centre for Islands, established at the Ministry of Maritime Affairs in 1994, functions today as the Islands Directorate of the Ministry of Regional Development and EU Funds, where it is hoped it will remain. In the meantime, it has been moved eight times from one ministry to another, as a centre, then a directorate, or a sector within a directorate, and even a department of a sector within a directorate. The fourteenth Government of the Republic of Croatia, which took office in 2016, returned it to the level of a directorate in 2017.

Since the 1999 act had undergone three rounds of amendments (OG 149/1999, 22/2002 and 33/2006), the first task of the department/sector, now the directorate, was to prepare a new Islands Act. When this was done, it seemed to have been forgotten that the act defined the implementer and authorised them to implement island policy, so that it was necessary to start again at the beginning and first define the policies, principles, aims and necessary authorisation; however, it turned out that procedural errors did not matter much, since the survey conducted on the islands in 2017, on a sample of 602 respondents, showed that the principles and aims set out in the NIDP 20 years earlier were still current and worthy of implementation. Questions were asked about each principle and aim, and the answers were affirmative, in a range of 62 to 77% (Institute of Economics, Zagreb, and Ministry of Regional Development and EU Funds 2017). Most affirmative answers, however, were accompanied by comments saying that no one had benefited, as the NIDP had been long forgotten, and the 1999 act only implemented in part.

The newly established directorate, with a new deputy minister, could be content with this, as the programme had never been deactivated, and there were people in the directorate with 5, 6, or even 15 years of experience in implementing the act. In addition, the Strategy for Regional Development of the Republic of Croatia up to 2020 had recently been adopted (OG 75/2017), along with the Strategy for Spatial Development of the Republic of Croatia (OG 106/2017), and development strategies for all six counties that cover inhabited islands were also under preparation. Notably, the Strategy for Regional Development introduces the category of areas of developmental particularities, including the islands, and recommends the introduction of an index of sustainable island development. Although it was mentioned less frequently, the Act on the Adoption of the Protocol on the Integrated Management of the Coastal Area, derived from the Barcelona Convention, was no less important. It includes all Mediterranean islands in the coastal area (OG International Treaties 8/12). It was considered that enough principles and aims had been defined, and work began immediately.

A group of employees and external consultants noticed at once that their work would be easier than that of the long-disbanded group that had produced the first draft of the act in 1999. This was because Croatia had already been a member of the EU for three years and had respected and accepted almost everything the EU proposed. Among other things, it was necessary to respect everything in terms of the islands, and this soon proved to be of benefit. At that time, the EU was beginning to abandon the attitude that islands were so diverse that their development should be left up to the countries to which they belonged. Step by step, a European island policy began to be formed. The first network of island authorities (ISLENET) was formed by 1993. It operates under the political umbrella of the Islands Commission of the Conference of Peripheral Maritime Regions (CPMR), promoting sustainable energy and environmental management. In 2007, the EU Parliament adopted the Motion for Resolution, which proposed tailored measures for European islands. In 2011, a political initiative for European islands, called the Pact of Islands, was officially recognised by the European Parliament. It has been growing ever since and today includes 117 members. Its most important document, the "Resolution on the Special Situation of Islands," was adopted by the EU Parliament in 2015. It called for an Agenda for EU Islands, accompanied by an EU Strategic Framework for Islands, with a view to linking financial instruments that could have a major territorial impact. The next year, the Smart Islands Initiative, a bottom-up group of European island authorities and communities, produced the Smart Islands Declaration. These campaigns, achievements by the EU Parliament and strategic documents (the chronology is best presented by Smart Islands Initiative, 2016) have attracted attention, and the islands have gone from being a neglected region scattered throughout the EU to a subject of European development policies in less than 10 years.

Croatian representatives at the EU Parliament and other EU bodies arrived in Brussels in 2013, and quickly got involved in everything that had to do with islands. Five years of tortuous preaccession negotiations proved to have been a good training ground: From the outset they participated actively in almost every island initiative and were particularly active in setting up the Intergroup for Sea, Rivers, Islands and Coastal Areas. The intergroup was established in 2015, bringing together 81 members of the European Parliament (MEPs) from 20 different EU member states. One of the most active MEPs when it comes to islands is a Croatian. He has been elected vice chairman of the intergroup.

Meanwhile, back at home, in the spring of 2017, a working group for preparing a new draft Islands Act was faced with many old and new strategies and programmes. Anyone who opposed the new act was given the EU documents and the Constitution of the Republic of Croatia to read. The work progressed faster and more smoothly than in 1999, and the new act was brought to the table in 2018. It was adopted just before Christmas of 2018 (OG 116/2018).

A STEP FORWARD?

In the public debate before the adoption of the act, it was claimed that the new act introduced some new elements but basically remained the same. *Plus ça change, plus c'est la même chose* was the most often heard remark. The key provisions will remain unimplemented. There are more subsidies and discounts than before, but they will be approved as before. Vertical coordination depends on LSUs that will always ask for more funds than could be earmarked in the state budget and beseige the Islands Directorate and other state institutions. There was also talk of too few signs of improved horizontal communication between ministries and public companies, so the directorate would continue to be invited to the openings of new bridges, clinics and water supply systems, built by others, but without its knowledge.

A comparison of the old and the new, however, shows that there is more room for improvement in island development policies than ever before. During the drafting of the new act, it was confirmed that instutional capacities are greater than before, and it was shown that, partly because of the awareness of the islands as a region requiring special policies, and partly because of the obligations acquired upon accession to the EU, there was more political will to implement them. Besides, comments were received from the northern, most developed islands that had not been heard in 1999, when they claimed, "The state administration is slow and hinders us from managing our own development," "Help us by not interfering" and so on. Neoliberal economists pounce on such things with great satisfaction but should not forget that when shoving the state aside, they often conceal the development agendas of large investors, megaprojects that damage the environment and affect the local community, which is hardly acceptable on the mainland, let alone on the islands. It should not be forgotten that a considerable number of islanders do not oppose such investments, claiming at the same time that they do care about their island's carrying capacity. The topic deserves much more attention than we can devote to it here, but the message that should be heard by everyone is quite clear: Development policies for near islands must be balanced in terms of mainland (top-down) and island (bottom-up) components.

The new act is better balanced than the old one and introduces principles and aims that were not in the 1997 NIDP. For the first time in a legal act, the notion of islandness appears, defined in the introduction and later used to explain several development measures. Islandness is defined as a "set of geographic, social, historical, economic and ecological specificities resulting from being completely surrounded by the sea" (Article 5). A sentence has been added that apparently cannot be avoided in documents about near islands: "As an exception, the notion of islandness also applies to the Pelješac peninsula." The intention to follow EU policies is obvious in another new addition. In the same article, for the first time in island legislation, a definition of a "smart island" is given. It is an

> island which, by using appropriate tools and innovative solutions, is developing in an ecologically, socially, technologically and economically sustainable way, building a circular economy and increasing its self-reliance and resistance to climate change: A smart island is a component of overall spatial-economic development policies, since it can be used as a template for planning and promoting the development of sustainable technologies which, when tested and introduced in island circumstances, may be adopted in greater dimensions on the mainland. (Article 5)

Both definitions are used in articles on development measures when subsidies or incentives need to be justified. There are explanatory notes: "respecting islandness," "due to pronounced islandness," "sustainable island development is stimulated and supported by projects implemented in line with one or more of the guidelines for smart islands" and so forth.

Critiques of the act in public debate have mostly concentrated on development documents that, following the logic of strategic planning, should have preceded the adoption of the act. The act stipulates the preparation of a National Island Development Plan, to be adopted by the Government of the Republic of Croatia, at the proposal of the ministry. The National Plan will define the implementation of aims in the National Development Strategy that relate to islands and must be aligned with spatial planning documents at every administrative level. It is adopted for a period of seven years. Island Development Plans adopted by island LSUs are also stipulated. These are to be prepared in accordance with the programmatic areas of the National Plan and serve as a basis for financing development projects using EU funds and other sources. Like the National Plan, the island development plans cover a period of seven years. According to the new act, Island Annual Programmes must be prepared in line with the National Plan and island development plans. Annual Island Programmes are adopted by the minister by 30 September at the latest in the current year to become part of the debate on the state budget for the next year.

Critics of the new act did not fail to notice the similarities between the NIDP in force from 1997 and the newly stipulated National Island Development Plan, and the similarities between the never-adopted SIDPs and newly stipulated Island Development Plans. The Island Annual Programme from the old act has been revived under the same name with just one difference—it is not adopted by Parliament but by the minister competent for island affairs. On the other hand, critics admitted that the act has aligned these programmes and strategies, and that the procedure has been significantly simplified. In addition, island development plans are now compulsory, if they wish to compete for EU funds, which means there will be an end to allocating resources at the whim of island mayors.

The section of the act that caused the most heated debate was about *island rights*. Islanders and island NGOs who participated in the public debate listed the benefits, stimulatory measures and subsidies affirmed in the old act, and claimed these as *vested rights*. What is more, they demanded that they applied to not only islanders living on the islands, but also those who have moved to the mainland and reside on the islands occasionally. In the debate, individual interests that claimed to be of a general nature clashed with interests of a state that had to have the power to adjust measures to conduct development policy efficiently. Observations that some subsidies and incentives should not be introduced at the level of the *acquis communautaire* were attacked from all sides, and at the end of the day, politics overcame development policies. A section was inserted in the act called "Island Rights," which stipulated subsidised prices in maritime, air and road traffic, free bridge crossing and subsidised water costs if the island was supplied by water conveyance. To this end, special island identity cards would be issued. Details on issuing and using these IDs would be provided in a separate law, so the debate continued in the same acerbic tone.

In contrast to the 1999 act, the new act was more selective and better defined the subject of development policies. It was high time to more closely define the position of inhabited islands that belonged to local self-government units with seats on the mainland or neighbouring islands. Such islands only have local boards and, in the NUTS classification, are defined at the lowest LAU 2 level. There are 31 of them, or to be more precise, 27 (since four of them did not even have local boards). The islanders of these 31 islands had representatives in the bodies of units of local self-government, but since they come from sparsely inhabited islands and are few, their influence on voting or the formulation of policies is negligible. They do not have their own budgets but are awarded budget lines by their LSUs over which they again have little influence. They do not even have their own bank accounts and so have to seek authorisation for the payment of any expenses. Lawyers' opinions on whether they could even be considered as legal persons differ.

The islanders of Male Srakane and Vele Srakane, for example, have their local board on the nearby island of Susak and have to go there and ask Susakians to pass their request to the town of Mali Lošinj. Susakians then add their own requests and go to Lošinj for the outcome, which is rarely in anyone's favour. The inhabitants of Biševo do not have a local board at all and must go to Komiža on the island of Vis, where they can apply to the town administration, not as legal representatives of their island, but as individuals with requests of public interest. Separated as they are by the sea from the system to which they formally belong, the few, elderly inhabitants of such small islands have so far felt no benefit at all from the principles outlined in the NIDP and insufficiently addressed in the old Islands Act. In 20 years of implementation of the island policy, small islands have not became "equal parts of Croatia in terms of development policy." Furthermore, the principle according to which an island is a "development unit whose development is the primary concern of islanders" can sometimes be seen as an example of bureaucratic irony, particularly on islands with an average age of older than 70. It should be added that legislation regarding local and regional self-government, which has mushroomed from 1992 to the present day, allows local self-government units to delegate some of their responsibilites to local boards, which would ensure their legal standing and ability to manage their own finances, bid for tenders and so on. Since the adoption of the Act on Local Self-government (OG 90/1992) until now, there have been no recorded instances of self-government units on islands making any such gestures.

The drafters of the new act nonetheless remembered the principles of the NIDP on the "active role of the state and county administration and local self-government in island development." Small islands, including those with some inhabitants, although official statistics record them as uninhabited, are listed in a special article of the act as "islands with specific status" (Article 10). For these islands, the ministry adopts the Programme of Measures within the Island Annual Programme to "mitigate the shortcomings which arise from the specific status of these islands in relation to the seat of local self-government within which they fall" (Article 10). Speaking of gestures, it should be noted that the initiative to distinguish small islands as a separate issue did not come from island or coastal-island LSUs, but from the Islands Directorate. It has been demonstrated that a top-down, bottom-up policy can only succeed if the initiative comes from both sides and that if there is no bottom-up element, those at the top must take full responsibility.

One reason why the issue of small islands was not tackled earlier has already been mentioned—lack of island data. Official state statistics touch the small islands only every 10 years, during population censuses, and during the other nine, data are kept at the LSU level. For the first time the new act stipulates a system of island development indicators that should cover all inhabited and, to a lesser extent, uninhabited islands. In Article 12, which is pretty ambitious, it is stated that the level of development of an island is evaluated by means of island development indicators, respecting islandness; that the purpose of such evaluation is to establish the state and direction of development on each island; that the evaluation procedure is to be conducted by the ministry responsible for islands and that state administration bodies and public companies are obliged, at the request of the ministry, to supply the information necessary for evaluation and monitoring the degree of development on each island. The system has many indicators and data coming from all sides that must be entered promptly and in the right place for the directorate to gain an accurate, multidimensional insight into what is lacking and what should be given priority. An institutional analysis, which was not skipped over this time, showed that the state administration had enough staff for these tasks and was sufficiently equipped in IT terms. It was also shown that it would rarely need to gather primary data, as this had already been done in many places, and that the main problem would be collecting what was there and discerning what was significant for the islands and what needed to be put into the system. As in 1999, the implementers have a great deal of work in front of them.

CONCLUSION

Is this a step forward? Has the supposedly improved new act established mainland-made island policy, influenced by the equally mainland-made policies of the EU, and will either be successful? The question requires some analysis, as there is no consensus on the meaning of "successful" here. The nearer an island is to the mainland, the easier it is to deinsularise it, and the development of individual Croatian islands and neighbouring mainland so far has shown that this happens without specific island policies. If the bearers of island policies find this successful and turn to building bridges and tunnels, and providing more frequent fast ferry lines and air connections, will they know how to go about deinsularisation while retaining islandness at the same time? Will they realise the reason for this in time, because islandness is not only heritage, but also an asset?

In the period between the two acts, an entire movement sprang up that saw the small islands as living laboratories for development and testing new technologies. And indeed, it is asked, where else could this happen, for is it not on the small islands that the skills of sustainable use of resources have been mastered, skills that now need to be shared and developed? New sustainable solutions should be tested in island conditions until they are ready for upscaling and implementation on the mainland. Smart cities, however large, start with small smart islands.

The new Croatian Act has identified this kind of island development, defined what is to be done and, in many places, emphasised that inevitable deinsularisation must be accompanied by the preservation of islandness, using it as an economic potential. The story has begun well. Let us hope it will not end with the deinsularisation of all near islands, their incorporation in mainland networks and their islandness abandoned. The Croatian island policy would then be merged with an overall regional policy, which would assume deinsularisation to have taken place, and therefore cease all consideration of island specificities. Timely warnings about the value of islandness could and most certainly will be issued as there are enough island knowers and policy analysts in Croatia ready to consult the state policymakers. They could convincingly explain that the deinsularisation policy can result in win-win solutions. Another question is, will the other side listen?

REFERENCES

Braudel, Fernand. 1972. *The Mediterranean and the Mediterranean World in the Age of Philip II*. New York: Harper & Row.

Čuka, Anica, Lena Mirošević, Josip Faričić, and Vera Graovac Matassi. 2017. "Phylloxera Revisited: The Spread of Grapevine Disease in Dalmatia and Its Influence on Socioeconomic Development and Agricultural Landscape." *Annales, Series Historia et Sociologia* 27, no. 1: 101–18. https://doi.org/10.19233/ASHS.2017.09.

Institute of Economics, Zagreb. 2018. "Analiza gospodarstva s projekcijama i scenarijima gospodarskog razvoja Primorsko-goranske županije do 2030. godine." Zagreb: Institute of Economics, Zagreb.

Institute of Economics, Zagreb, and Ministry of Regional Development and EU Funds. 2017. "Analitička podloga za izradu novog Zakona o otocima—završno izvješće." Unpublished manuscript.

Lastovski statut. 1994. Split: Književni krug Split.

Ministry of Regional Development and EU Funds. 2018. "Izvješće o učincima provedbe Zakona o otocima u 2017. godini." *Ministry of Regional Development and EU Funds*, https://razvoj.gov.hr/o-ministarstvu/djelokrug-1939/otoci/zakon-o-otocima-nn-116-2018/izvjesce-o-ucincima-provedbe-zakona-o-otocima-u-2017-godini/3786 (accessed 24 May 2019).

Smart Islands Initiative. 2016. "Smart Islands Initiative." *Smart Islands Initistive*, http://www.smartislandsinitiative.eu/en/index.php (accessed 26 May 2019).

Starc, Nenad et al., ed. 1997. *National Island Development Programme*. Zagreb: Ministry of Development and Construction of the Republic of Croatia.

Stulli, Bernard. 1959. *Ustanak mornara u Boki Kotorskoj 1.—3. februara 1918*. Split: Izdanje mornaričkog glasnika.

World Bank. 2016. "Croatia Policy Notes 2016." *World Bank*, http://documents.worldbank.org/curated/en/876601468000628285/pdf/103938-WP-P156959-PUBLIC-Croatia-Policy-Notes-2016-EN.pdf (accessed 26 May 2019).

Chapter Eleven

An Outlier

The Island of Susak

Julijano Sokolić and Nenad Starc

The island policies described in the previous chapter have proved to be policies of deinsularisation implemented according to the act that explicitly advocates the preservation of islandness. The two are not mutually exclusive, but state budget investments in the islands show that deinsularisation is a more important goal than preserving islandness. Investments in the infrastructure linking the islands with the mainland far surpass investments in preserving the intangible island heritage. They also outweigh investments in technologies that need to be promoted on the islands along with the *smart* concept. Until now, there have only been minor investments in pilot installations using smart technology (desalination, harnessing solar energy and so on), which are being tested for potential later use on other islands or the mainland, in *smart* towns. It is clear that small and large islands are treated differently, as demands for connections with the mainland are much louder and better heard when they come from large islands with higher populations, which accommodate more tourists and find themselves overwhelmed by transport problems every summer. The small islands, which are more suitable for testing *smart* technologies, as they are more insularised and the islanders are more skilled at using scarce resources, are left to the side.

The islands have been called "canaries in the mine," because they are the first to suffer if the measures of island development policies do not have the expected effect. Such fragile islands do exist in the Adriatic. In the Croatian archipelago, there are many among the small outer islands that, as has already been stated several times, are stagnating economically and losing their populations, while the larger ones are gaining, right under the noses of the bearers of island development policies. They manage as best they can, in different ways, and it is not easy to single out one island that represents them all. In the Croatian archipelago, there are many islands that deviate from the general picture of karst, vegetation and rocky shores; therefore, we need to first look at those that deviate the most, which is somewhat harder.

Each have their own particularities; some are near the mainland, and others are presumptuously called open-sea islands. Most, of course, are small or very small: volcanic Brusnik and Jabuka; Palagruža, which is nearer to the Italian coastline than to the Croatian one; Unije, with its loess field; Silba, the island of seafarers; and several others. If we consider these islands and all of their specificities, and compare them to one another, from the geology and flora to the economy and history, one small, sandy island in the northern Adriatic stands out more than the rest. Perched on the rim of the archipelago is Susak, also known as Sansego, Sansig and, long ago, Sansagus.

It covers only 3.75 km^2. The nearest mainland coastline in Istria is 25 nautical miles, and the nearest island, Lošinj, is 6 nautical miles away. Sitting on the rim of the archipelago, it really does feel like an open-sea island and is often considered in that light; however, the wider picture of the Adriatic and still-wider Mediterranean prove that it is in fact a near island. How it came to be there is still the subject of scientific jousting. This hill of sand in the middle of the sea is easier to describe than explain. According to some, in the Ice Age, when the sea covered only half of the modern Adriatic, the rivers deposited sand on the uneven ground where there were karst plateaus. Some of these plateaus had clay beds, which gradually collected sand and formed mounds. According to others, the deposited sand was picked up by the wind, sieved and blown to the plateaus. Yet, others think the rivers and winds were equally responsible. And there are those who think everything began with a volcanic eruption (Gračanin 1957).

What is known is that at some point ice melted, the seas rose and the Adriatic doubled in size, reaching the area that would later become the Venice archipelago. Thus, along with the typically Mediterranean karst islands, a sandy island, Susak, also appeared 10,000 years ago in the Adriatic, a geological conundrum, an island that does not match the rest. On the other islands, the hands of the islanders have created an anthropogenic landscape, and it would take a fire to find the vineyards and olive groves of yesteryear beneath the macchia. Susak is also anthropogenic but to the extent that we can merely guess what it must have looked like 2,500 years ago when the first islanders arrived. At one point, much later, Susakians turned it into a vineyard, and the population swelled to 1,760, almost 470 people per km^2, more than on any other Adriatic island, and more than on any other small Mediterranean island. The story of how this happened, like the island itself, is special.

ON THE RIM OF THE ARCHIPELAGO

Susak was settled in the Illyrian era. Little is known about the small island, and the farther back in time we go, the less there is to find. We can surmise that the Romans drove the Illyrians away, and the Ostrogoths drove out the Romans as they fell back toward Byzantium. In 884, when the name Sansagus was first mentioned, Susak was still under the cloak of Byzantium and, for a little while, under the Franks. Venice was growing and gaining in strength at the time, and, from the 9th century onward, tried and eventually succeeded in overpowering the eastern Adriatic coast and its islands; however, this was not without resistance and wars, as from 910 onward, Croatian rulers were also encroaching on the Adriatic. For a time, they succeeded in exacting levies from Venice for the right to sail toward Otranto, the exit from the Mediterranean. Fighting for the throne weakened them, which Venice took advantage of in the 10th century, ceasing to pay the levies and starting the second millennium with war.

With the consent of Byzantium, Venice conquered the coastal towns and strongholds, along with almost all the islands, including Susak, in the year 1000. Centuries of rivalry began, when the rulers of the islands changed over and again, depending on the fortunes of war and the relationships between the great powers. Croatian King Krešimir II, notable for taking back the coast and islands for a period, gave Susak to the Benedictine order in 1025, and the island gained permanent owners and inhabitants. The king also found the means to build a monastery for monks to pray and work in solitude (Turčić 1998). Pirates sailed the Adriatic in those days, so the monastery was not built on the shore, but on higher ground in the island interior. The interior is also less windy, even today. Susak cannot be reached if there is a strong *bura* or *jugo* wind.

Susak thus became an ecclesiastical fief and remained so until the end of the 19th century. At first it belonged to the monastery and later to a parish in the Osor Diocese. Smaller islands are harder to find in archives than larger ones; however, we do know that the monks invited laypeople to come to the island and that they indeed responded (Turčić 1998). We do not know where they came from or how many there were—probably as many as was sustainable with all things considered. Nevertheless, a farming leasehold known in the Mediterranean as the *colonat* (see Chapters 3, 6 and 10) was established on Susak. It was also a prevalent mode on other islands; the newcomers could leave if they chose, but better prospects were not visible on the horizon. Thus, for centuries, they decided time and again to stay on land as tenants. It was not theirs, but they could make a living and produce surpluses for their landlords. They made themselves at home, became islanders and even answered the question about how the island came into existence. The legend is well known in the archipelago.

Long, long ago, a stone plate resting on four pillars rose out of the sea, as we were told by Anton Picinić-Toninov, a devoted storyteller and the chairman of the Susak local board on a summer evening some 30 years ago. As time went on, the wind blew enough sand to form an island, while storms gradually dislodged the four pillars. One broke, and then another, until the island was resting on only two. Storms eventually wore away at these, too, and that is why when the *jugo* lashes or the *bura* rages, you can hear groaning from the bowels of the island as the sand shifts and falls. Storms are fearfully anticipated, because if another pillar breaks, the entire island might sway and topple into the depths of the sea. The people of Susak do not like the *bura* or the *jugo*, and they detest a high sea. Nonislanders of today who decide to disembark for Susak and cannot land because of stormy weather should reflect on whether the first newcomers ever wondered how such a small, loose pile of sand could possibly resist the sea.

We do not know why the Benedictines began to abandon Susak, but when the last monk left in 1770, there were enough inhabitants that remained and considered themselves indigenous (Sokolić 1993). The Church was considered in the same light and did not relinquish the gift received from so long ago; the island got a priest, then a parish church, and the tenants remained tenants. Changes in rulers did not mean changes in the Church. The great powers that possessed Susak were all Catholic, and the monks and later the parish priest welcomed those governing them without fear and bade them farewell without any major upsets. Even the Great Schism of 1054 left no trace on the island. The boundary to the east where the Orthodox Church territory cut through the Adriatic was distant enough from Susak. It stayed on the western, Catholic side; the tributes remained the same and it disrupted none of the labouring and fishing in the daily lives of the islanders.

Ora et labora applied to everyone; the Benedictines followed the basic rule of their order, and the people of Susak, the rules imposed on them. They reserved a third of their income and crops for the monastery (later the diocese). The skills they developed showed that they perhaps prayed less than the monks, but they certainly worked more. Although they could row to the mainland, and often did, the people of Susak could only watch the merchant galleys that sailed by without docking and the warships that did not attack them. The island community was too small to be significantly included in events across the sea.

In light of the eternal search for generators of growth and "locomotives of development," economists cannot fail to notice that the island owners took their third without wondering how to increase the production from which it came. They also cannot fail to notice that in such circumstances the island community could not develop a competitive edge and that no comparative advantages existed. Close to the geographical centre of Europe, and only a day's sail from Venice, Susak remained on the margins, and the people of Susak developed an almost self-sustained island economy. Rare visitors could report on the small village at the island peak and its poverty, as well as the skills the islanders had mastered and the efforts they invested. Susak could not support beasts of burden, and therefore everything had to be carried on the head and shoulders.

The islanders learnt from a young age how to tend the vines, level and stabilise the terraces on the sandy slopes, produce their own agricultural tools, make wine, weave nets, maintain boats, sail and fish, salt fish, forecast the weather, make clothes and footwear, carve wood, do carpentry and cook, all with careful management of food and fuel. The island shores are rocky, but the soil is sandy. The sandy layer is as much as 100 metres deep, but there is also black soil and other deposits that are fertile and almost everywhere arable. The parcels were all alike and uniformly fertile, and conditions for using and working them were equally analogous, which granted a similar life, and equal survival, for everyone. The Susak population grew faster than those on neighbouring islands, and the island soon became the most densely populated in the archipelago. This should not surprise us. This island, the Adriatic exception, almost all of which can be transformed into plantations of anything and everything that thrives in a Mediterranean climate, is capable of feeding more mouths than the others.

Susak and its people had lived in this way for hundreds of years. Under the wing and employ of the Church, they observed how the Croatian kingdom passed and the Hungaro-Croatian Kingdom followed, how Byzantium benevolently relinquished the Adriatic to Venice and how this only fuelled the fight for domination. After Byzantium and Venice left the stage, they witnessed the expansion of Austria and for a short time felt a slight air of bourgeois society brought in by Napoleon and taken away again by the Habsburgs. After the departure of the French, everything returned to normal. The people of Susak continued to work for themselves and others, and the tribute remained the same, as did its collectors. It was not until the late 19th century, in 1874, to be exact, that the already well-established Austro-Hungarian authorities passed and implemented a special act on redeeming land (Mirković 1957). It only applied to the March of Istria and nearby islands, and it allowed the tenants to purchase the plots of land they had worked. After long centuries of working for others, Susakians were allowed to buy their parcels in 20 annual instalments and remain on their own land.

The repurchase ended tributes to the Church, and much smaller taxes were introduced. Susakians continued to work their land, although now as proprietors with full control of land use. At that time, a large naval base was springing up in Pula, a small town at the tip of the Istrian peninsula, 34 nautical miles west of Susak. Not quite eight nautical miles to the east, boatbuilding and seafaring were thriving on the island of Lošinj. Susakians were far from being able to compete with them, but they were close enough to row across and appear on the Pula and Lošinj markets with their salted fish, table grapes and, above all, wine. The demand rose steadily, and Susakians responded as best they could.

They paid off their instalments, moved down from the village on the island peak to the largest cove and began building another village (known as *donje* [lower], as opposed to *gornje* [upper]). Their quality of life improved and the population increased. In 1843, there were 988 of them, and in 1869, according to the first official state census, there were 1,095 islanders (Sokolić 1993; Turčić 1998 and Korenčić 1979). Nearby accessible markets changed the economic structure of the island. The autarchic agriculture that had produced cereals, fruit and vegetables, along with grapevines, started to disappear under new vineyards. When Pula and Lošinj filled up with sailors and boatbuilders and shipowners, respectively, Susakians also began to abandon fishing. The island, which could survive despite the mainland, became entangled in a network of mainland economic flows and began to specialise in only one activity.

Specialisation meant being permanently present on the markets of other islands and the mainland as sellers or buyers. On the markets of Lošinj, Cres and Pula, islanders also appeared from nearby Ilovik and Unije, but none were as regular as the people of Susak. Accordingly, familiarity with Susak grew greater than that of other islands or the mainland, which has lasted until the present day. The traditional costumes of the Susak males, which they distinguished, developed and supplemented as they knew best, did not differ greatly from those of other islanders and the people of Istria. The Susakians were instead identified by their archaic speech and fine wines; however, the Susak women did differ at first glance. Their short skirts, which did not cover the knee, made them distinct in the entire Croatian archipelago and Mediterranean. They were also renowned for something far more indecorous. On such a small, remote island, they could not avoid inbreeding.

There were marriages between second and sometimes first cousins (Dolinar-Osole 1957), which had become normal throughout time, and the island became the target of rumours in the archipelago and a place of interest for the medical profession and anthropologists. In the 1950s, when there were still about 1,400 islanders on the island, it was established that the Susak children were no different from the Lošinj children in terms of their skeletal growth (Brožek, Ratko 1957) and that women entered menarche and menopause at the normal time and did not have pathological pelvic features (Kralj-Čerček 1957; Mikić 1957; Vižintin-Dekaris 1957). Anthropometric measurements showed that both women and men were rather tall, well-built and healthy (Škerlj, Mikić and Dolinar 1957). Researchers spent two years on the island and reported that in the middle of the Adriatic, 34 miles from the mainland, content, realistic, simple people were living, adhering to traditional morals.

Research into mental deviations, the results of which are usually greeted with scepticism in insular environments, displayed a rather different picture. It turned out that 9.2% of the people of Susak were psychiatrically remarkable and that their diagnoses included oligophrenia, senile dementia and schizophrenia (Lopašić and Mikić 1957). The family trees of the Susak families did not indicate incest as the main cause of these problems, but it was shown that they were not more prone to them than comparable communities. The structure of their illnesses was, however, different (relatively more cases of oligophrenia and schizophrenia), so Susak was specific in terms of certain mental illnesses and their frequency (Lopašić and Mikić 1957). This aroused wider interest, so further anthropological studies were arranged using more advanced methods and instruments.

It was noticed that Susak was the most distant genetic isolate, not only among the islands, but also in Croatia as a whole, and that this had led to genetic consequences. In the meantime, there were fewer and fewer Susak islanders. The samples taken by researchers were less and less representative, so that at the beginning of this millennium, when less than 200 native Susakians were still living on the island, the sample was barely large enough to form any conclusions. Similar island isolates were sought throughout the world, and the most similar one was found in the middle of the Atlantic. The inhabitants of Susak and those of the island of Tristan da Cunha were found to have surprising genetic similarities (Bohaček 1964; Šarac 2017). A self-evident question relates, of course, to the proximity of the mainland. On Tristan da Cunha, it is not easy to leave the island in search of marriage partners, and no one visits the island for that purpose. But it was easy enough to row from Susak to the mainland. Thus, the question remains as to why wives and husbands were not sought in other places. The answer to such a question is usually sought in proprietary relationships, size of parcels, inheritance issues and so on. In the case of Susak, it seems it was simply another specificity.

PHYLLOXERA

At that time, 2,000 km away in Europe, changes that would be felt in the Adriatic decades later, mostly of ill fame, were beginning. In the late 1850s, English botanists brought vine seedlings from America and planted them in their vineyards. They grew well, and cross-breeding and new wines were contemplated, although no one could even guess that the roots came in infested with a dangerous pest. Through ignorance and confusion, it was given many Latin names: *Phylloxera vastatrix*, *Dactylosphaera vitifolii* and *Peritymbia vitifolii*. In Croatia it was also known as vine lice, root-eater, root-biter or simply "filoksera," a blight that is still talked about on the Croatian islands more than any past wars or earthquakes.

The home of phylloxera was the Mississippi Valley, where American vines had grown for centuries. Those vines resisted and adapted; their roots thickened, hardened and became resistant, so the vines and their pests coexisted. European vines, however, were caught unaware. The pest ravaged the few English vineyards and managed to cross the Channel in 1865. It occupied the Rhone valley and continued its onslaught, destroying about 70% of the French vineyards by 1895. It raged throughout Spain and Italy, and, in 1880, reached the vineyards around Zagreb in Croatia. In 1886, it appeared in Istria, and in the early 1890s, it cropped up on the Kvarner islands. It continued south and, by 1925, had blighted all the Dalmatian islands (Čuka, Mirošević, Faričić and Graovac Matassi 2017).

At the turn of the century, it reached its peak in Europe, and the number of places where vineyards still flourished could be counted on one hand. One of those islands was Susak. Phylloxera did not completely pass it by and was noticed later than in other places in 1905, but it never took hold (Medarić 1957). The sandy soil on which the vines had thrived for centuries proved too acidic for the pest and even came to be known as antiphylloxeric soil. The vineyards remained healthy, displaying the colours of the endemic Susak grape varieties every autumn. News of thousands of hectares of ruined vineyards and hundreds of French vintners who had committed suicide may have reached the island of Susak, but the phylloxera blight did not become etched in the collective memory as on the Dalmatian islands. What did reach them on a daily basis, as it had before 1905, was news about the price of wine. And it was very good news.

After centuries of self-sufficiency, which was only threatened by the demand for fish, table grapes and wines on the markets of Pula and Lošinj, the people of Susak discovered almost overnight that they held a comparative advantage. Suddenly, their sandy soil seemed boundlessly exploitable, and none of them hesitated to turn their parcels of wheat, broad beans or figs into vineyards. By 1903, 87% of the island was covered by vines, and an island that was already so exceptional in other ways now became the exception in a vine-growing archipelago, and indeed the entire coastal area. Repurposing the land continued, and the spacing between vines grew smaller, as did the number of varieties. Terraces were formed on the slopes, the soil was enriched by hoeing and catch crops were passed over. By 1908, the vines had reached the first houses in the village. In the end, there were about 1.8 million vines, covering 96% of the island. Susak was transformed into a big vineyard, an island with a monoculture (Medarić 1957).

Susakians grew grapes and made wine to sell, and bought everything else they needed on the mainland. Everyone knew how to plant and tend vines, so the daily routines did not require much change. Vine-growing intensified to such an extent that other jobs were reduced to a minimum, and the people only went fishing in the absence of vineyard tasks. Although this near island had only been important to its own islanders for centuries, now it became important to others, thanks to what it could offer to the mainland, which could be reached by rowing. It did not become highly significant, of course, because it was still small. Vintners and traders on the mainland did not cast their eyes toward Susak; by their own strength the Susakians created a small, although important, hub in the local and regional network of wine production and sales. It was small but big enough to do everything itself, from production to sales. A golden opportunity had been seized with both hands.

"A man's vineyard possesses him," so goes the island saying. It demands time and effort, so the people continued to work hard and long, and made money; however, mutual cooperation was uncommon. The *colonat*, which ended when the land was redeemed, left behind small, scattered parcels and a multitude of owners. At the time, and indeed later, no islander had enough income to buy a piece of land larger than 10 hectares. There were no buyers from the mainland or larger islands, and Susak remained densely crisscrossed with boundaries between small parcels, limiting their owners to small-scale commerce. A half-century later, in 1953, there were still 358 rural homesteads on the island, of which 216 were less than one hectare in size, and only one was slightly larger than eight hectares. Almost 90% of the grapes and wine went for sale, and each person sold his own produce (Medarić 1957). On this densely populated island, people hoed and pruned shoulder to shoulder, but each homestead made wine in their cellar, carried the barrels to the boats they owned or hired and, if necessary, sailed themselves to Pula or Lošinj.

Susakians were skilled vineyard workers, like all islanders at the time, virtuosos with their mattocks, with which they hoed or deepened the vineyards on the sandy slopes. But like all islanders, they were mediocre when it came to their cellars. They lacked knowledge of the vintner's craft and the proper tools, so the wine was easily spoiled. They were also inapt when it came to trading on the markets, but still, sales of grapes and wine were good. Demand was constantly higher than supply, but Susakians did not go to market with large quantities of produce, they did not compete with one another and no one saw any reason to form a cooperative to reduce costs and make transport easier. The first cellar where wine was produced collectively was built decades after the monoculture was established on the island. The one and only wine-producing cooperative was founded after the Second World War (Jelaska 1957). Meanwhile, the population was growing. In 1880, there were 1,111 islanders. In 1890, there were 1,333 (Korenčić 1979). They needed more housing, so they built more dwellings in the lower village. To this end, they filled in a shallow cove in the island's main bay, so as not to use up precious vine-growing land.

Although it had seemed at first mysterious and apparently invincible, phylloxera was cornered. Grafting the European vines onto the resistant American roots began to yield results, and the crisis in Europe and the Adriatic began to recede. After it peaked and had affected vineyards in different locations between 1895 and 1905, production began to rise. By the outbreak of the First World War, the devastated French and Italian vineyards were mostly back on their feet. Phylloxera ravaged Adriatic vineyards even after the First World War, and islanders could not wait to learn how to graft the vines; in the 1920s, most of the vintners had left for Australia and both Americas. Shielded by the antiphylloxeric soil of their island, Susakians continued selling wine on the stable markets of Pula and Mali Lošinj. Each vineyard and cellar owner decided what and how much to produce within the family circle, the price of wine remained high and 96% of the island was still covered in vineyards. On this small island, overinhabited by smallholders familiar with vines and wines but not used to assessing the mainland markets, no one could have guessed what was happening beyond the sea and what it would mean to them. The skills known as strategic planning remained unknown on the small near island. On the mainland, no one seemed concerned about the vineyard in the middle of the sea.

At the same time, the great plan the Austrian navy started implementing in the late 19th century proved beneficial. Besides building fortified harbours and assembling the fleet, Austrian strategists aimed at safe navigation in the Adriatic (Stulli 1959). The rim of the archipelago became important, and attention turned toward the outer, mostly small islands, including Susak. Regardless of the people of Susak and their wine, and perhaps not even knowing that they had not succumbed to phylloxera, the Admiralty included the island in their plans. A small harbour was built in the main island bay in 1874, and a beacon was lit on the highest peak in 1881. The lighthouse, whose light shone

19 nautical miles out to sea, was finished in 1886, and manned by a telegraph and lighthouse crew (Stulli 1959). Being on the rim of the archipelago meant getting a powerful lighthouse. From then onward, the island could be seen from the mainland and formed part of a network, with news arriving fast over vast distances. Thirty years later, telegrams arrived from St Petersburg with news of the October Revolution. They were sent to everyone, from Pula to Boka Kotorska, and the sailors in Boka mutinied on 1 February 1918, calling others to join them (Stulli 1959). On that day, a red flag flew from the Susak lighthouse. It was lowered the next day, but the island was no longer completely on the margins of world events.

Although they were now better connected and informed, the people of Susak did not seem to realise that they were no longer counted among the rare few who could produce wine. They continued to tend their vines and even started to exhaust their plantations. The last agronomic analyses produced in the early 1950s discovered domestic vines grafted onto American stock in more than half the vineyards (Medarić 1957; Jelaska 1957). The reason was not phylloxera, but yield. Susakians had noticed that grafted vines produced more fruit and that the first yield in a new vineyard would quickly appear. As far as fertiliser was concerned, they had never had enough natural fertiliser on the island. They purchased it on the mainland and brought it back on the boats that ferried wine to markets. When artificial fertilisers reached the island, the people welcomed them with open arms, but it soon became clear they were using them more than was necessary. The vines grew fast, which was easy enough to see, but the grapes did not ripen properly and the wines were rather poor. This was also obvious, but sales were good, and there were seductive goods to be bought in Pula and Lošinj. As they began to accumulate more things, they needed more income for purchases; therefore, they continued to produce as much as possible in the shortest possible time, without allowing the vineyards to rest, without composting them and without catch crops to enrich the soil. Vineyard specialisation threatened sustainable vine-growing. It turned out that the imperative of sustainable management of island resources disappears when an island is included in wider mainland economic flows.

The last agronomic analysis confirmed that the people of Susak, who responded swiftly when the price of grapes or wine rose, responded grudgingly when it fell. In 1952, a half-century after the monoculture was established, 82% of the island was still covered in vineyards, and production was still intensive (Medarić 1957). Phylloxera had been eradicated long ago and consigned to the collective memory, and wine was no longer an exclusive European product, arriving from both Americas, Africa and Australia. Prices were not as high as previous years, and tourism was starting to mushroom on the Adriatic coast and islands, creating a demand for other products that thrived in the Susak sandy soil. Susakians, however, who had showed that they could exploit a comparative advantage when it appeared, did not know how to react when it faded away.

A SMALL ISLAND AMONG BIG PLAYERS

During the half-century of monoculture, Susak survived four states. The winds of two world wars blew over it. After the collapse of Austro-Hungary at the end of the First World War, the Kingdom of the Serbs, Croats and Slovenes disembarked on the island. By 1920, Susak had been abandoned to Italy. After the capitulation in 1943, the Italians withdrew, and the Germans moved in for a short time; then at the end of the Second World War, the newly established Democratic Federation of Yugoslavia appeared on its shores. After dispensing with all political parties except the Communists, it was renamed the National Federative Republic, and later the Socialist Federative Republic of Yugoslavia.

Susak stayed on the sidelines of major disembarkations and destruction, and could only judge the geopolitics of the great powers by the flags that were brought and flown on the island. As states came and went, Susak changed camps, too—three times between 1918 and 1920. When it was an Austro-Hungarian possession, it was on the side of the Axis powers, but in 1918, it was told it had gone over to the Entente, while in 1920, as a new Italian possession, it found itself against those who had liberated it in 1918.

The people of Susak are said to be cheerful folk by the people of Lošinj. Perhaps, some 150 km from the bloody front on the River Soča, they watched with bitter smiles as the war tragedy reflected on their small island as a farce. They were not, however, spared the adversities of war. With their trade interrupted, and on the verge of famine, they reverted to cereals and vegetables, because there were always more hungry mouths to feed in wartime than in peacetime. Susak was no exception; throughout history, all the Croatian islands have been havens for war refugees and defectors.

Whether it was for old times' sake or because they simply did not know otherwise is hard to say, but the people of Susak kept reverting to their vineyards and wines. When the markets reopened and the refugees left, the cereals and vegetables again had to make way for the vines, and it started all over again. The vineyard monoculture that had kept going through two world wars began to collapse under the socialist economy and the onslaught of development measures that caused more harm than good to the islands, particularly the small ones. By the late 1950s, a new chapter in the history of Susak was being written.

HOBOKEN, NEW JERSEY

Closer to the geographic centre of Europe than Rome or Vienna, Zagreb or Belgrade, Susak was again left on the margins, even though it had again been caught up in an economic net cast from the mainland. Susak survived the vine-growing and wine-making net of European dimensions thanks to its sandy soil and skills. It was now entangled in the net of socialist industrialisation, where it was doomed to perish, as would soon be demonstrated.

The economic policy of so-called primary accumulation of capital (see Chapter 10), which pressurised agricultural workers into accepting lower than market prices, did not bode well for an island that had once specialised in vine-growing and wine-making. For decades, Susakians had been involved in networks of market exchange, and now they were expected to sell their wine at low prices to the state stores. The most important source of income for many Susakians began to dry up. In the middle of a planned economy, the people of Susak started to plan their departures. Mostly, they went across the ocean, because they thought they had nothing to expect on the domestic mainland. This time, circumstances did not smile on them.

On the one hand, during the war, Susak had been a refuge and had 1,760 inhabitants in 1945, the highest number in its history (Korenčić 1979). On the other, in spite of good years, it was no longer possible to make a living from grapes and wine, and the fish that might have helped them were destined for canneries in the archipelago. By the time of the 1948 census, there were 1,629 permanent inhabitants on the island. Five years later, there were 1,434 (Korenčić 1979). They mostly migrated, illegally, to the United States. Young lads would join the crews of merchant ships and then jump ship in the first American port. Or they would row to Italy, go to the mainland and cross the border illegally on foot, seeking asylum in Italy or Austria. The island from which they had travelled to sell their wines for decades was suddenly not so easy to leave. During the 1950s and 1960s, only the most entrepreneurial spirits left—about 400 of them.

The strictness of socialism, however, began to ease. In the mid-1960s, Yugoslav authorities decide to open the exit doors just a crack, and then fully. It was possible to get a passport through regular channels, and at last people were allowed to travel abroad more than once. The easiest, most frequent destinations were European countries in search of workers. Many people became guest workers, and small villages on the mainland in poorly developed areas began to lose their populations. The even smaller villages on the small islands faced an exodus. By 1971, according to the official census, there were only 323 inhabitants left on the island of Susak.

People had been leaving the Croatian islands via the mainland or ocean since the late 19th century. Due to supposed opportunities for a better life, or just because others were going, and sometimes because they heard stories that life was better across the ocean, yet always reluctantly and with sorrow and expectations, they arrived in both Americas and Australia, New Zealand, South Africa and elsewhere. The push factors, as economists call them, were always stronger than the pull factors. The islanders would leave in groups, sometimes large ones, cross an ocean, sometimes two, and arrive in places where no one expected them. Once becoming accustomed to their new surroundings, they would split up and begin searching for fellow countrymen. They formed associations and clubs, and met during patronal festivals of their native islands. Susakians left, too. The island had always been overpopulated, and some had left earlier, between the two world wars.

During the long Susak evenings, Toninov used to tell us that they went to the United States just before the Second World War, crossing the continent to the West Coast and finding work on the border between the United States and Canada, in the area between Seattle, Tacoma and Vancouver. They worked as lumberjacks, in sawmills, and in timber transport. Many used the skills they had brought with them and turned to fishing. After the Second World War, maritime traffic between the United States and Europe boomed, and there was plenty of work in the ports on the East Coast. Now accustomed to looking for work, they went east again and, for reasons no one knows, chose the little town of Hoboken on the Hudson River in the state of New Jersey.

In the 1960s, when they began to arrive there in larger numbers, Susakians showed yet again that their island was special. In contrast to many other dispersed islanders, they went only where they would find fellow islanders and quickly established an enclave where they spoke "the Susak way" and in many ways lived as they had done on their island. In Hoboken, they took care of new arrivals from Susak, helping them find work, buy houses and even set up stores where the customers did not need to speak English. The Sansego Soccer Club was founded with its own premises for talking soccer and other things. For years, dances were held on Saturdays, only for Susak folks. The youth would dance, and the grandmothers and aunts would line the walls and indulge in matchmaking.

Andrija Matešić, a dedicated Susakian, patiently recorded all this for years, so the story of the island that went to America is supported by numbers. He told us that in the late 1980s, there were no fewer than 2,500 Susak people living in Hoboken, New Jersey, and nearby Fairview, more than the highest population ever recorded on the island itself. Matešić told us that he was not much surprised, as women from Susak continued to marry only men from Susak, and men from Susak would only propose to Susakians. In Hoboken, one could come into the world as a native Susakian. The topic found its way into literature. In 1997, a novel about Susak was published in Zagreb. Its plot revealed the duality that has marked the lives of Susakians in the United States and even expressed the wish they know will never come true: an American woman with Susak heritage visits the island, falls in love with one of the rare young Susakians and stays. Entitled *Rhapsody on the Island of Susak* (Basarić 1997) it was soon translated into English and sold out in Hoboken and Fairview (the second edition sold out, too).

The gathering places today are still St Anne's Church in Hoboken and the Church of St John the Baptist in Fairview. There is also the St Nicholas Society of Sansego, the Susak *klapa* singing group. The Susak women gather in the Society of the Holy Rosary. Susak is not special in this regard. Croatian islanders have always founded part of their identity in their faith; however, it is quite remarkable how much the people of Susak are connected by their faith, and how much it has helped them and still helps them to retain their islandness in the American melting pot. They also keep up connections with the Croatian Heritage Foundation, and Pino Morin Skulcinov, a native of Susak, stood for mayor in Hoboken in the late 1980s. He did not win. Although, in the Sansego Soccer Club, they agree that it probably helped his business.

THE THIRD MILLENNIUM: A HOLIDAY ISLAND

When they moved to America, the people of Susak took their island with them. There are many streets in Hoboken where Susak speech can be heard, customs are kept and surnames survive, although children are given both Susakian and American names. The soccer the Croatian emigrants take with them wherever they go is still played. They soon realised, however, that they could not move their island entirely, and that Susak of New Jersey and Susak of Croatia will remain apart.

One of the most important activities of the Sansego Soccer Club and its president, Bruno Morin Skulcinov, is organising group visits to the home island. In 1967, for the first time since the end of the Second World War, 160 American Susakians paid a moving visit, which they all wanted to repeat. Since then, it has happened every year. Skulcinov told us that it has quickly become a habit to spend one's annual holiday on the old island. In 1968, the Hoboken visitors could only stand by and observe as yet another mainland net almost trapped the island. In that year, the INA state oil company brought drilling equipment to the island and started drilling in the middle of the soccer field in the lower village. The oil derrick was 46 metres high, taller than anything ever built on the island, and drilled down to a depth of 5,000 metres (Sokolić 1993). Nothing was found, so they packed up and left, without even asking whether it was appropriate to arrive without previous notice and start drilling 50 metres from the first houses. Accustomed as they were to the sudden appearance of comparative advantages, as well as uninvited landings, this time Susakians could only make it a topic of their evening gatherings in the Sansego Soccer Club.

As their visits increased in frequency, the Americans, as émigré islanders are known on the Croatian islands, including Susak, were able to observe how the infrastructure systems from the mainland began to reach the island. Slowly but surely, 43 years after the arrival of the telephone cable, an underwater cable that connected Susak with the state electricity network was laid in 1979. Much later, in 1993, the post office in Susak acquired a VHF connection with the automatised telephone exchange on the island of Krk and four public phone booths, while the capacity of the telephone exchange on the island was increased to 256 connections. There was no mobile network, so the new connection was seen as a major achievement. In the winter of 1979, only three months after the arrival of electricity, a hurricane-level *jugo* devastated the harbour wall and waterfront, sinking many boats and flooding the lower village. There was a great deal of damage that took years to repair, but the news reached Hoboken quickly. There was much serious talk in the Sansego Soccer Club about the Susak underwater pillars and the suspicion that they might one day founder.

The real estate business was launched at that time, however, proving that buyers interested in empty houses on Susak thought that the island was absolutely stable and attractive. After almost 20 years of hesitation, longer than on other small islands, the islanders started to sell houses to "weekenders," people from the mainland who renovated them and started to use them as holiday homes, mostly in the summer, since the island was too far from the mainland for weekends. Every summer since then, the remaining 200 or so natives of Susak have welcomed about 800 second-home owners, along with more and more Croatian and foreign tourists. In the autumn, they revert to their secluded, everyday lives on the island, which once had 470 souls to every square kilometre. In 1988, there were 684 house numbers on Susak. We were told in the island local board that there were about 100 permanently inhabited homes, about 250 second homes and a similar number of houses waiting to be renovated and sold. That same year, renovation work began on the parish church and lasted two years. The Hoboken Susakians collected $128,000 for that project.

In the early 1990s, both islands perked up. On both Susak and in Hoboken, the independence of Croatia was eagerly watched, and almost everyone joined the Croatian Democratic Union (CDU), the triumphant party in the democratic elections. The war that followed raged from 1991 to 1995, far from the island, as in the past. Susakians are long-lived, and those who were born before the First World War could boast that they had survived through five states without ever leaving their island. The CDU did not forget the little island that had given it great support; so far three presidents of the Republic of Croatia have visited it and made all sorts of promises, quite a few of which were fulfilled. Many photographs have been taken during those visits, particularly presidential selfies that today adorn the largest wall in the Susak Futsal Club in the lower village.

Immediately after the establishment of the new state in 1991, on the ground stretching southeast from the graveyard, land for a new, modern vineyard was levelled. This time, the investor was an Italian with Lošinj roots. In 1991, 32,000 seedlings of various varieties were planted. The project had been launched during Yugoslav days, when the doors gradually opened to foreign investors, and it continued to find a welcome in independent Croatia. The necessary permits were acquired quickly, but the state administration could not cut through the overgrown, once carefully tended paths to the vineyard. This was done by Susak people from Hoboken. The Picinić quartet, Ivan/John, Vito, Stefano and Nikola, set aside $40,000 and bought a bulldozer, which they delivered to the island in 1989. That year, as usual, they arrived on holiday and taught a local man how to dig the paths using the bulldozer, dispose of the waste and be of service to the islanders. Until then the priest had needed to ask the people of Lošinj to come to Susak when there was a death in the lower village because he could not gather six able men on the island to carry the coffin along the steep, impassable path to the graveyard.

Two years later, new seedlings were brought via the newly cleared path to the vineyard. That same year, 188 inhabitants were recorded as living on the island. The new investor found it hard to raise a workforce, and when the 2001 census revealed that there were still only 188 inhabitants, it became clear that the bulldozer had cleared the path only for funerals. The vineyard, where as many workers as were needed once would have worked, would have to remain unattended. Wine production did not entirely fail, and the wine from endemic Susak black grapes had reasonable sales; however, serious attempts at renewal failed. The 2011 census did not inject any optimism, as only 151 inhabitants were counted on the island. According to the islanders themselves, there are no more than 130 permanent inhabitants today.

On the other hand, the islanders must admit that the island infrastructure is much better equipped. In 2001, the water supply and drainage system began to be built, with a reservoir of 750 cubic metres, and in 2012, the pupils of Susak Elementary School were included in the e-learning system of the base school in Mali Lošinj. The main road was paved in 2016, and in 2017, a desalination plant began operations. It is large enough to ensure a constant water supply to the reservoir even in summer, when tourists and holiday homeowners swell the island population 10 times over. The new jetty, consisting of a harbour wall and ferry dock, which is a satisfactory 147 metres long, were completed in 2018. Intangible culture has not been overlooked. The traditional procession in honour of the Great Cross was restored in 2017, and the archaic Chakavian Susak local speech, which was dying out in competition with that of newcomers from the mainland, has been protected as an intangible cultural good of the Republic of Croatia.

A measure implemented in 1998, by the town of Mali Lošinj, was particularly promising. Supported by approbation from Hoboken, the people of Susak established a home for the elderly and incapacitated, and the town provided a building it owned for this purpose. It was expected that the Americans might return to their island in their old age and that the home would always be full. During the first couple of years, a few old people who had always lived on the island went to the home, while just a few advance reservations were received from Hoboken. It was not enough, and to the surprise of many, the home closed its doors in 2012. Infrastructure projects that were widely considered at that time and later implemented were envisaged for the young and old alike. In the following years, however, they were actually carried out only for second-home owners and guests, and islanders who were still far from elderly or incapacitated. Judging from the last two censuses, most of these are newcomers, new islanders with their own view of the island and what should be done on it. The old, native Susak folk could only watch as the key was turned in the lock of the home and conclude that the island had finally moved to Hoboken.

After 2,000 years of inhabitation, wars and rulers; the use and lack of use of comparative advantages; labouring and finally moving to another continent, Susak stepped into the third millennium. Susakians in the New World did not find a new island, but a small harbour town by a megapolis, where they settled in and developed and doubled their numbers, thanks to their mutual care and solidarity. They copied and pasted the island, and preserved it as best they knew. Fifty years later, however, it had been consigned to their memories, and they continued to live as real Americans. The Susak that they left in the Adriatic retained too few islanders to cherish it as in the past, and the people on the mainland did not care for it as they should. It seems that the departure of 90% of the population in the past 30 years has not really upset anyone except for the islanders themselves. Special in so many ways, in this respect Susak is no different from other small Croatian islands.

Today, as in the past, it is burdened by the antinomy of a small near island, insularised yet caught every so often in a net cast from the mainland from which it can barely escape. The last one, woven by the construction of infrastructure for islanders who are no longer there and heralded as a welcome achievement of island development policies, is meant primarily for temporary, summer users. The remaining native islanders can only wonder whether development requires something else and what the purpose of the pipes and cables, shipping lines and fast internet really is, if at the same time they do not encourage activities focusing on the sustainable exploitation of the island's resources. With better infrastructure than it has ever had, Susak has fewer inhabitants than ever and is gradually turning into a holiday island. The remnants of the community that once produced much and lived hard lives now accommodate and feed tourists, produce almost nothing and live easier lives than ever. Determination to go on has not weakened, however, even among the elderly who could make a contribution to the discussion on what constitutes islandness. They have a serene view of birth and death as coming into the world (i.e., the island) and setting out to sea when the time comes. We cast off and move on, they say.

Anton Picinić-Toninov cast off for the last time some 20 years ago. He remained optimistic to the end. When we chatted that evening 30 years ago, he finished his story about the shaky island by looking up to the starry sky. "I do not fear drowning," said the man who had tied up his boat so many times on Susak and other jetties. "If the last pillar broke, the island would sway a bit and then moor itself to the Universe." There was no reply to that, but he indicated that he had not finished. "We will last," he said, "we will last."

WILL THEY LAST?

The question, of course, relates to the islanders and islandness, although it can also be asked in the context of climate change. The rising of the sea level by a metre would have a definite effect on Susak, as it would submerge the rocky base on which the island sand lies, reach the vineyards and cause considerable losses through erosion. The harbours would also be endangered on almost all the Adriatic islands, as the difference between neap and ebb tides is small, so the coastal villages have been built right on the shore. There are more frequent occurrences of bad storms such as not have been recorded previously, cutting off communications with the mainland and prohibiting small boats from going out to sea. The answer that the islanders, particularly those on the small Croatian islands, have always managed to master the forces of nature, and will continue to do so, no longer stands.

An insight into the demographic history of the small islands reveals that none of them have more than 10% of the population maximum that was achieved between the two world wars. There are simply no islanders who can cope with climate change, and those who have remained too often declare that their island is sinking. We heard this on Olib and Susak, which have their own chapters in this book, and also on the other small islands. Among that 10% there are also newcomers, new islanders who are only just building their islandness. It cannot be expected that they will earn it immediately, in the first generation. Pessimists would say that haste is not necessary; if newcomers are reduced to being the caretakers of summer islands during the winter months, they will not need traditional island knowledge and skills to welcome summer-residence owners each summer and report that everything is in order.

Of the 47 inhabited Croatian islands, 28 have been officially classified as small. Optimists will say that they will remain islands, even if most are reduced to summer resorts. We agree with this. To whatever extent they are only inhabited during the season, they will still be suitable places for testing environmental technologies and remain open to new settlers, new islanders who are prepared to build a new islandness. The history of the skilled islanders and sustainable use of scarce resources does not need to end as the illegal mushrooming of holiday homes suggests. The Adriatic islands are near islands, and their history is filled with those who have taken refuge there or settled permanently, before and after the arrival of the Croats. Whoever came to stay became an islander, and there is no reason why this should not happen again. This time, however, they will need help. Building a new form of islandness needs to be emphasised and assisted top-down and bottom-up, and all stakeholders must be involved. The previous chapters have shown that this is not unfeasible and that it is possible to construct the necessary institutional framework for directing island policies.

Pessimists will say that this is the easy part but that implementation has shown that the bearers of the island policy are unable to coordinate all the actors. They will also mention the lack of political will. Optimists will reply that a lack of coordination is typical of all development policies and can also be found on the mainland and in countries with much better developed state administration systems, so that this should not be a disincentive. They will appeal to the rising number of *smart* initiatives that have been conceived and implemented by the islanders themselves. Again, the pessimists will have a retort to this, and so the debate will go on.

This kind of debate, which is of course useful and increasingly common in Croatia, highlights the stakeholders who should not be allowed to remain insularised. We refer, of course, to nissologists and policy analysts, whose contribution to managing island development may be of crucial importance. So far they have observed and researched the Croatian islands well and contributed with their analyses and insights into island development. Their work is available to the bearers of island policies. The question remains, however, whether they should get involved directly as active stakeholders in preparing and implementing island programmes, development plans, training courses and so on. There are various answers to this question, but everyone agrees on one thing: The role of the "mine canary watchers," however scientifically useful, must be rejected in advance.

REFERENCES

Barlović, Mladen. 1957. "Čeljusti i zubi." In *Otok Susak*, edited by Mijo Mirković, 561–67. Zagreb: JAZU.

Basarić, Zdravko. 1997. *Rapsodija na otoku Susku*. Zagreb 1997.

Basioli, Josip. 1984. *Ribarstvo na Jadranu*. Zagreb: Nakladni zavod Znanje.

Bognar, Andrija, Ferenz Schweitzer, and Gyula Szoor. 2003. *Susak: Environmental Reconstruction of a Loess Island in the Adriatic*. Budapest: Hungarian Academy of Science.

Bohaček, Nenad. 1964. "Tristan da Cunha and Susak." *Liječnički vjesnik* 86, no. 11: 1,412–16.

Brožek, Josef, and Buzina Ratko. 1957. "Rast djece i krvno srodstvo." In *Otok Susak*, edited by Mijo Mirković, 389–97. Zagreb: JAZU.

Čuka, Anica, Lena Mirošević, Josip Faričić, and Vera Graovac Matassi. 2017. "Phylloxera Revisited: The Spread of Grapevine Disease in Dalmatia and Its Influence on Socioeconomic Development and Agricultural Landscape." *Annales, Series Historia et Sociologia* 27, no. 1: 101–18. https://doi.org/10.19233/ASHS.2017.09.

Dolinar-Osole, Zlata. 1957. "Prilog poznavanju genealogije." In *Otok Susak*, edited by Mijo Mirković, 459–75. Zagreb: JAZU.

Gračanin, Zlatko. 1957. "Pedološke prilike." In *Otok Susak*, edited by Mijo Mirković, 73–113. Zagreb: JAZU.

Jelaska, Marcel. 1957. "Lozne sorte i njihova obrada." In *Otok Susak*, edited by Mijo Mirković, 237–61. Zagreb: JAZU.

Korenčić, Mladen. 1979. *Naselja i stanovništvo Socijalističke republike Hrvatske 1857–1971, Djela JAZU knjiga 54*. Zagreb: JAZU.

Kralj-Čerček, Lea. 1957. "Utjecaj prehrane, tjelesne građe i socijalnog porijekla na vrijeme pojave menarhe." In *Otok Susak*, edited by Mijo Mirković, 435–40. Zagreb: JAZU.

Lopašić, Radoslav, and Fedor Mikić. 1957. "Duševne poremetnje." In *Otok Susak*, edited by Mijo Mirković, 535–51. Zagreb: JAZU.

Medarić, Jura. 1957. "Vinogradarska monokultura." In *Otok Susak*, edited by Mijo Mirković, 201–36. Zagreb: JAZU.

Mikić, Fedor. 1957. "Menarha i menopauza kod Suščanki." In *Otok Susak*, edited by Mijo Mirković, 441–43. Zagreb: JAZU.

Mirković, Mijo. 1957. "Otočna zajednica Suska." In *Otok Susak*, edited by Mijo Mirković, 3–42. Zagreb: JAZU.

Šarac, Jelena. 2017. "The Remote Isolate of the Croatian Littoral: Fifty Years of Research on the Island of Susak with New Insights into Its Genetic Diversity." *Collegium Antropologicum* 41, no. 1: 11–18.

Škerlj, Božo, Fedor Mikić, and Zlata Dolinar. 1957. "Prosječni odrasli Suščanin." In *Otok Susak*, edited by Mijo Mirković, 443–55. Zagreb: JAZU.

Sokolić, Julijano. 1993. "Otok Susak—mogućnosti revitalizacije." *Društvena istraživanja* 3, no. 4–5: 12–13.

Stražičić, Nikola. 1975. "Cresko-lošinjska otočna skupina." *Otočki ljetopis Cres-Lošinj* 2: 143–90.

Stulli, Bernard. 1959. "Ustanak mornara u Boki Kotorskoj 1.—3. februara 1918." Split: Mornarički Glasnik.

Turčić, Anton. 1998. *Susak, otok pijeska, trstike i vinograda*. Susak: Župni ured Susak.

Vižintin-Dekaris, Nada. 1957. "Ginekološki nalazi." In *Otok Susak*, edited by Mijo Mirković, 559–60. Zagreb: JAZU.

Appendix A

The Islands of the Croatian Archipelago

Inhabited Island	Island Area (km^2)	Island Coastline (km)	Settle-ments	Population in 2011	Population Density per Km^2	Households	Age 0–19	Age 20–59	Age 60+	Largest Settlement (Inhabitants)
Kvarner Islands										
Krk	405.2	180.6	68	19,383	47.8	7,824	3,410	10,572	5,401	Krk (3,730)
Rab	86.1	121.0	8	9,328	108.3	2,892	1,785	4,964	2,579	Rab (437)
Cres	405.7	268.2	30	3,079	7.6	1,265	528	1,713	838	Cres (2,289)
Lošinj	74.7	112.2	5	7,587	101.6	2,778	1,451	4,484	1,652	Mali Lošinj (6,091)
Ilovik	5.5	14.1	1	85	15.5	36	14	36	35	Ilovik (85)
Unije	16.9	38.1	1	88	5.2	47	13	34	41	Unije (88)
Vele Srakane	1.2	7.4	1	3	2.5	2	0	3	0	Vele Srakane (3)
Male Srakane	0.6	3.9	1	2	3.3	1	0	0	2	Male Srakane (2)
Susak	3.8	12.9	1	151	39.7	62	22	69	60	Susak (151)
Total	—	—	116	39,706	—	14,907	7,223	21,875	10,608	—
North Dalmatian										
Pag	284.2	302.5	25	9,059	31.9	3,590	1,674	4,545	2,840	Pag (2,849)
Vir	22.1	31.9	1	3,000	135.7	1,320	477	1,506	1,017	Vir (3,000)

Ugljan	51.1	78.8	7	6,049	118.4	2,281	900	2,903	2,246	Kali (1,638)
Ošljak	0.3	2.4	1	29	96.7	16	2	13	14	Ošljak (29)
Pašman	60.1	70.2	8	2,845	47.3	1,048	506	1,391	948	Tkon (763)
Vrgada	2.3	9.2	1	249	108.3	113	30	90	129	Vrgada (249)
Olib	26.1	33.3	1	140	5.4	75	5	50	85	Olib (140)
Molat	22.2	51.6	3	197	8.9	123	3	38	156	Molat (170)
Rivanj	3.6	10.4	1	31	8.6	23	0	8	23	Rivanj (31)
Sestrunj	15.1	29.4	1	48	3.2	37	2	14	32	Sestrunj (48)
Zverinac	4.2	14.3	1	43	10.2	26	0	13	30	Zverinac (43)
Iž	16.5	35.2	2	615	37.3	328	53	217	345	Veli Iž (400)
Rava	3.6	16.0	1	117	32.5	66	6	27	84	Rava (117)
Silba	14.3	26.2	1	292	20.4	154	23	134	135	Silba (292)
Premuda	8.7	25.7	1	64	7.4	40	4	20	40	Premuda (64)
Ist	9.7	23.0	1	182	18.8	95	1	61	120	Ist (182)
Dugi Otok	113.3	182.1	11	1,655	14.6	704	208	729	718	Sali (740)

(continued)

Inhabited Island	Island Area (km^2)	Island Coastline (km)	Settlements	Population in 2011	Population Density per Km^2	Households	Age 0–19	Age 20–59	Age 60+	Largest Settlement (Inhabitants)
Murter	17.6	42.6	4	4,895	278.1	1,943	932	2,409	1,554	Murter (2,025)
Prvić	2.4	10.6	2	403	167.9	200	33	173	197	Prvić Šepurine (239)
Zlarin	8.1	20.2	1	284	35.1	144	25	113	146	Zlarin (284)
Krapanj	0.4	3.6	1	170	425.0	68	27	83	60	Krapanj (170)
Kaprije	7.1	25.2	1	189	26.6	103	10	73	106	Kaprije (189)
Kornat	32.4	68.8	—	19	—	12	1	16	2	—
Žirje	15.1	41.8	1	103	6.8	66	2	24	77	Žirje (103)
Total	—	—	77	30,678	—	12,575	4,924	14,650	11,104	—
					Central Dalmatian					
Drvenik mali	3.4	12.0	1	87	25.6	66	0	29	58	Drvenik mali (87)
Drvenik veliki	11.7	23.9	1	150	12.8	99	4	41	105	Drvenik veliki (150)
Čiovo	28.1	46.7	6	5,908	210.2	2,254	1,160	3,056	1,692	Okrug Gornji (3,081)
Brač	395.4	175.1	22	13,956	35.3	5,312	2,522	7,248	4,186	Supetar (3,213)

Šolta	58.2	79.5	8	1,700	29.2	829	189	753	758	Grohote (449)
Hvar	297.4	270.0	27	11,077	37.2	4,254	2,009	5,719	3,349	Hvar (3,771)
Vis	89.7	84.9	16	3,445	38.4	1,411	582	1,729	1,134	Vis (1,672)
Biševo	5.9	18.2	1	15	2.5	10	0	8	7	Biševo (15)
Total	—	—	82	36,338	—	14,235	6,466	18,583	11,289	—
South Dalmatia										
Šipan	16.2	29.4	2	419	25.9	169	77	204	138	Šipanska luka (212)
Lopud	4.4	14.6	1	249	56.6	97	31	136	82	Lopud (249)
Koločep	2.4	12.9	1	163	67.9	65	22	89	52	Koločep (163)
Korčula	271.5	190.7	10	15,522	57.2	5,442	3,163	7,985	4,374	Vela luka (4,137)
Mljet	98.0	135.2	14	1,088	11.1	469	170	539	379	Babino Polje (270)
Lastovo	40.8	49.0	5	792	19.4	286	156	408	228	Lastovo (350)
Total	—	—	33	18,233	—	6,528	3,619	9,361	5,253	—
FULL TOTAL	—	—	308	124,955	—	48,245	22,232	64,469	38,254	—

Sources: *Census of Population, Households and Dwellings, 2011*; Croatian Bureau of Statistics, Zagreb, www.dzs.hr; *Statistical Yearbook 2018*, Croatian Bureau of Statistics, Zagreb, www.dzs.hr; Duplančić Leder, T., T. Ujević, and M. Čala. 2004. "Coastline Lengths and Areas of Islands in the Croatian Part of the Adriatic Sea Determined from the Topographic Maps at the Scale of 1:25,000." *Geoadria* 9, no. 1: 5–32.

Appendix B

Tourism on the Croatian Islands, 2010–2017

	2010		2017	
Island	*Arrivals*	*Nights*	*Arrivals*	*Nights*
Cres	99,927	646,546	122,616	883,357
Krk	137,920	759,433	215,620	1,274,168
Rab	138,599	1,020,431	181,254	1,298,089
Lošinj	—	—	—	—
Unije	1,001	6,650	—	—
Susak	—	—	493	4,445
Ilovik	271	2,515	32	301
Pag	99,901	717,596	113,474	808,893
Vir	51,117	366,895	77,366	598,204
Premuda	45	727	—	—
Silba	3,032	27,929	3,165	27,560
Olib	—	—	193	2,058
Molat	288	2,533	370	3,363
Ist	—	—	479	4,151
Iž	—	—	1,833	16,527
Rava	—	—	120	1,246
Osljak	174	2,011	329	2,789
Rivanj	—	—	—	—
Sestrunj	—	—	—	—
Ugljan	8,236	74,927	10,961	94,868
Pašman	14,986	132,016	18,577	167,015
Vrgada	290	3,442	11	35
Dugi Otok	—	—	—	—
Zverinac	—	—	251	2,566
Prvić	2,509	19,927	1,825	13,128
Murter	38,612	202,109	39,923	299,921
Kornati	—	—	—	—

(continued)

Krapanj	—	—	22,394	160,877
Kaprije	—	—	1,150	8,994
Zlarin	—	—	8,895	37,714
Žirje	376	2,958	793	7,057
Čiovo	—	—	—	—
Drvenik Mali	—	—	1,538	14,063
Drvenik Veliki	—	—	808	7,494
Šolta	13,862	99,102	17,287	145,342
Brač	—	—	—	—
Hvar	86,216	439,909	197,361	713,292
Vis	18,134	119,446	28,090	162,881
Biševo	—	—	82	77
Koločep	—	—	5,528	42,131
Lopud	3,257	23,385	13,276	78,501
Šipan	—	—	—	—
Korčula	56,846	295,714	80,939	376,423
Lastovo	4,710	31,594	7,332	49,079
Mljet	14,698	70,002	30,637	134,471

Source: *Statistical Reports, Tourism in Seaside Towns and Municipalities 2010 and 2017*, Croatian Bureau of Statistics.

Index

www.ingramcontent.com/pod-product-compliance
Ingram Content Group UK Ltd.
Pitfield, Milton Keynes, MK11 3LW, UK
UKHW012036150926
475560UK00001B/40